OCCASIONAL
PAPER

# Protecting Commercial Aviation Against the Shoulder-Fired Missile Threat

James Chow, James Chiesa, Paul Dreyer,
Mel Eisman, Theodore W. Karasik, Joel Kvitky,
Sherrill Lingel, David Ochmanek, Chad Shirley

INFRASTRUCTURE, SAFETY, AND ENVIRONMENT

The research described in this report was supported through provisions for independent research and development in RAND's contracts for the operation of Department of Defense (DoD) federally funded research and development centers: RAND Project AIR FORCE (sponsored by the U.S. Air Force), the RAND Arroyo Center (sponsored by the U.S. Army), and the RAND National Defense Research Institute (sponsored by the Office of the Secretary of Defense, the Joint Staff, the unified commands, and the defense agencies). The research itself was conducted within RAND Infrastructure, Safety, and Environment (ISE), a unit of the RAND Corporation. The mission of ISE is to improve the development, operation, use, and protection of society's essential built and natural assets; and to enhance the related social assets of safety and security of individuals in transit and in their workplaces and communities.

**Library of Congress Cataloging-in-Publication Data**

Protecting commercial aviation against the shoulder-fired missile threat / James Chow ... [et al.].
p. cm.
"OP-106."
Includes bibliographical references.
ISBN 0-8330-3718-8 (pbk. : alk. paper)
1. Aeronautics, Commercial—Security measures. 2. Terrorism—Prevention. 3. Surface-to-air missiles.
I. Chow, James, 1966–

TL725.3.S44P76 2005
363.28'76—dc22

2004026555



Published 2005 by the RAND Corporation
1776 Main Street, P.O. Box 2138, Santa Monica, CA 90407-2138
1200 South Hayes Street, Arlington, VA 22202-5050
201 North Craig Street, Suite 202, Pittsburgh, PA 15213-1516
RAND URL: http://www.rand.org/
To order RAND documents or to obtain additional information, contact
Distribution Services: Telephone: (310) 451-7002;
Fax: (310) 451-6915; Email: order@rand.org

# Preface

Following the terrorist attacks of 9/11, the question of whether or not to install countermeasure systems to protect commercial airliners against shoulder-fired missiles has been an issue of vigorous debate among decisionmakers in the United States and abroad. This research effort was designed to inform the American public and decisionmakers regarding the potential utility of technical countermeasures and other policies that could help to protect commercial aircraft against attacks with shoulder-fired missiles. We examine operational, effectiveness, and cost issues involved with countermeasure systems. During the course of our study, we had access to contractor and government information related to countermeasures and threat systems; however, all of our results and conclusions are based upon publicly releasable or open-source information.

In the period immediately following the September 11, 2001, attacks on the United States, the RAND Corporation undertook several research projects relating to counterterrorism and homeland security topics as elements of its continuing program of self-sponsored research. This research is the result of one of those research projects. The work was supported through the provisions for independent research and development in RAND's contracts for the operation of the Department of Defense (DoD) federally funded research and development centers: RAND Project AIR FORCE (sponsored by the U.S. Air Force), the RAND Arroyo Center (sponsored by the U.S. Army), and the RAND National Defense Research Institute (sponsored by the Office of the Secretary of Defense, the Joint Staff, the unified commands, and the defense agencies). The research itself was conducted within RAND Infrastructure, Safety, and Environment (ISE), a unit of the RAND Corporation. This mission of ISE is to improve the development, operation, use, and protection of society's essential built and natural assets; and to enhance the related social assets of safety and security of individuals in transit and in their workplaces and communities. Dr. Richard Neu, Assistant to RAND's President for Research on Counterterrorism (at the time this research was performed), provided overall supervision for this work.

# Contents

# Figures and Tables

## Figures



## Tables

# Summary

Air travel has become an integral part of modern life. Terrorists have long understood this and have made commercial aviation one of their prime targets. Al Qaeda and its affiliates have both the motive and the means to bring down U.S. commercial aircraft with shoulder-fired missiles, also known as man-portable air defense systems (MANPADS). No such attempt has yet been made against a U.S. carrier, but given the measures being taken to preclude 9/11-style attacks, the use of MANPADS will unavoidably become more attractive to terrorists.

What might be done to prevent such an attack? We concentrate here on the capabilities and costs of onboard technologies to divert or destroy an attacking missile. Given the significant costs involved with operating countermeasures based upon current technology, we believe a decision to install such systems aboard commercial airliners should be postponed until the technologies can be developed and shown to be more compatible in a commercial environment. This development effort should proceed as rapidly as possible. Concurrently, a development effort should begin immediately that focuses on understanding damage mechanisms and the likelihood of catastrophic damage to airliners from MANPADS and other forms of man-portable weapons. Findings from the two development programs should inform a decision on the number of aircraft that should be equipped with countermeasures (from none to all 6,800 U.S. jet-powered airliners) and the sequence in which aircraft are to be protected.

If it is determined that U.S. commercial airliners should be equipped with countermeasures upon completion of the development program, they should be employed as part of a broader set of initiatives aimed at striking and capturing terrorists abroad, impeding their acquisition of missiles, and preventing them and their weapons from entering the United States. Attention should also be paid to keeping MANPADS-equipped terrorists out of areas adjacent to airports and improving commercial airliners' ability to survive fire-induced MANPADS damage.

A multilayered approach is important because no single countermeasure technology can defeat all possible MANPADS attacks with high confidence. Nonetheless, substantial protection can be achieved. Laser jammers, for instance, will be commercially available for installation aboard airliners soon and should be able to divert single or possibly dual attacks by the relatively unsophisticated MANPADS accounting for most of those now in the hands of terrorists. Ground-based high-energy lasers (HELs) intended to destroy approaching missiles could counter MANPADS of any degree of sophistication, but they are not ready for deployment in the next few years and have significant operational challenges to overcome. Pyrophoric flares used reactively offer the promise of a cheaper alternative with better potential to handle multiple attacks than laser-based systems, but their effectiveness at protecting large transport aircraft from any MANPADS attack is not well established, and they would be most likely ineffective against sophisticated future systems.

We estimate that it would cost about $11 billion to install a single laser jammer on each of the 6,800 commercial aircraft in the U.S. fleet. The operating costs of fleetwide countermeasures will depend on the reliability of the system. Extrapolating from early reliability data from the systems currently deployed on large military aircraft, the operating and support (O&S) costs for a commercial variant were assessed to be $2.1 billion per year for the entire commercial fleet. The full ten-year life-cycle costs (LCCs) for developing, installing, operating, and supporting laser-jammer countermeasures are estimated to be $40 billion. If reliability goals recommended by the Department of Homeland Security (DHS) can be achieved, the ten-year LCCs are estimated to be $25 billion.

When would such an investment be worth it? That is not a question answerable solely through quantitative analysis, but some light can be shed by four avenues of inquiry. First, what would be the likely economic costs of a successful attack? If we take into account the value of a lost aircraft and a conventional economic valuation of loss of life, the *direct* cost would approach $1 billion for every aircraft downed. The indirect economic damage from an attack would be far greater. These costs result from the loss of consumer welfare through preemption of a favored travel mode or reluctance to use it, as well as operating losses suffered by airlines subsequent to an attack. These amounts will depend primarily upon two factors: the length of any possible systemwide shutdowns in air travel and any kind of longer-lasting public reluctance to fly. Both factors are difficult to predict, but if air travel were shut down for a week (it was shut down for three days after 9/11), the economic loss would amount to roughly $3 billion during the shutdown itself. Extrapolating from the long-term effects of the 9/11 shutdown, losses over the following months might tally an additional $12 billion, for a total economic impact of more than $15 billion.

A second avenue of inquiry can help place the cost of MANPADS countermeasures in context. To what extent must homeland-security and other counterterrorism resources be expanded or diverted to fund this one effort to help respond to a single threat? The $2.1 billion annual O&S cost, should it be borne by the government, amounts to only about 6 percent of the annual DHS budget. The fraction is much smaller if the costs of operations in Iraq and Afghanistan are included in the base. However, the $2.1 billion is a substantial fraction of total current federal expenditures on transportation security.

Third, it must be recognized that loss of life and economic impact would not be the only costs of a MANPADS attack. The perceived inability of the U.S. government to prevent attacks on its citizens on its own soil would set back U.S. efforts to counter terrorist groups globally and could weaken U.S. influence across a range of other interests abroad. Such an attack would also cause unquantifiable losses of security among the U.S. populace.

Fourth, and lastly, while countermeasures have been demonstrated to be an effective resource in protecting our military aircraft, the circumstances of protecting commercial airliners from terrorists are sufficiently different that we should ask ourselves the following questions: Upon deployment of countermeasures, how easy do we think it will be for terrorists to adapt and find vulnerabilities to airliners through the use of weapons that are not affected by countermeasures? Would defenses against these weapons be possible, or would they require a similar level of funding to protect against?

A decision as to whether to proceed with a MANPADS countermeasure program must thus balance a variety of considerations. On the plus side:

- New countermeasure technology with capability against a variety of attack situations will be available in the near term, with the potential to avert the loss of hundreds or even thousands of lives and tens of billions of dollars.
- Funding such a system would require a reallocation or expansion of federal homeland-security resources of perhaps 5 percent—and a much smaller proportion of total federal counterterrorism resources.

On the minus side:

- Annual operating costs would represent nearly 50 percent of what the federal government currently spends for all transportation security in the United States.
- Well-financed terrorists will likely always be able to devise a MANPADS attack scenario that will defeat whatever countermeasures have been installed, although countermeasures can make such attacks considerably more difficult and less frequent.
- Installing countermeasures to MANPADS attacks may simply divert terrorist efforts to less protected opportunities for attack. To put it another way, how many avenues for terrorist attack are there, and can the United States afford to block them all?

Given the significant uncertainties in the cost of countermeasures and their effectiveness in reducing our overall vulnerability to catastrophic airliner damage, a decision to install should be postponed, and concurrent development efforts focused on reducing these uncertainties should proceed as rapidly as possible. The current DHS research, development, test, and evaluation (RDT&E) activities are a prudent step both toward reducing significant cost uncertainties involved and minimizing the delay of program implementation once a go-ahead decision is reached.

To summarize, any federal policy to protect against MANPADS should not be restricted to countermeasures development, but should involve multiple layers, with emphasis on the following areas:

1. Rapidly understanding and finding ways to reduce the O&S cost component of countermeasures in a commercial-airline setting. In addition, decisionmakers should be thinking about how specific countermeasure systems would work best in conjunction with other protection efforts and technologies. Understanding the weaknesses of countermeasures should help focus these efforts, and vice versa.
2. Focusing a concurrent technology development effort on understanding damage mechanisms and the likelihood of catastrophic damage to airliners from MANPADS and other forms of man-portable weapons such as rocket-propelled grenades (RPGs), mortars, and small-arms fire. This will serve three purposes: clarifying the damage caused by single or

multiple MANPADS hits on airliners, informing choices regarding the implementation of mitigating measures such as inerting fuel tanks and missile countermeasure systems, and assessing the seriousness of other forms of attack against airliners.

3. Working with international governments to slow down the proliferation of MANPADS technologies, in particular those against which countermeasures are less effective.
4. Putting together concepts of operation that integrate countermeasures into the overall aviation safety, security, and law enforcement system. These can help local law enforcement establish the size and location of airport security perimeters and define ways in which information from the onboard countermeasure system sensors can be used to help find, track, and apprehend MANPADS operators. Lastly, they would help provide an understanding of the costs from false alarms to air-traffic operations and local law enforcement.

# Acknowledgments

The authors would like to extend thanks to Richard Neu, Natalie Crawford, and Jim Thomson of RAND for their support and feedback during this effort. Other RAND staff providing valuable feedback included Bruce Hoffman, Russ Shaver, Brent Bradley, Tim Bonds, Tom Hamilton, Claude Berrebi, Charlie Kelley, and Richard Mesic. Discussions with numerous external parties provided a diverse set of inputs and greatly improved our understanding of the relevant issues. These parties included Patrick Bar-Avi (Rafael USA), Mike Barrett (NAVAIR), Christopher Bolkcom (Congressional Research Service), Richard J. Doubrava (ATA), Anthony R. Grieco, Elizabeth Guran (GAO), Yael Hiram (Rafael USA), Ernest "Burt" Keirstead III (BAE), Robert Kudwa (Kudwa and Associates), Igo Licht (IAI), Dave McNeely (Boeing), Alvin D. Schnurr (Northrop Grumman), David Snodgrass (Northrop Grumman), John Stanfill (Northrop Grumman), Jim Tuttle (DHS), and Col. Tal Yeshaya (Embassy of Israel). Following the original publication of this document, subsequent discussions with Mr. James Peppler at the MITRE Corporation were helpful in clarifying technical issues on countermeasures. Finally, all omissions or mistakes are the sole responsibility of the authors.

## List of Abbreviations

| | |
|---|---|
| AUPC | average unit production cost |
| $C^2$ | command and control |
| CG | command-guided |
| CLIRCM | closed-loop infrared countermeasure |
| CRAF | Civil Reserve Air Fleet |
| DHS | U.S. Department of Homeland Security |
| DIRCM | directed infrared countermeasure |
| DoD | U.S. Department of Defense |
| FAA | Federal Aviation Administration |
| FOC | full operational capability |
| FOV | field of view |
| FPA | focal plane array |
| FRP | full-rate production |
| FY | fiscal year |
| GDP | gross domestic product |
| HEL | high-energy laser |
| IR | infrared |
| IRCM | infrared countermeasure |
| IRST | infrared search and track |
| LAIRCM | large-aircraft infrared countermeasure |
| LAX | Los Angeles International Airport |
| LCC | life-cycle cost |
| LRIP | low-rate initial production |
| MANPADS | man-portable air defense systems |
| MEMS | micro electro-mechanical system |
| MTBF | mean time between failure |
| MTHEL | mobile tactical high-energy laser |
| MWIR | mid-wave infrared |
| MWS | missile-warning system |
| O&S | operating and support |
| PB | President's Budget |

| | |
|---|---|
| RDT&E | research, development, test, and evaluation |
| RF | radio frequency |
| RPG | rocket-propelled grenade |
| SAM | surface-to-air missile |
| SOF | Special Operations Force |
| UAV | unmanned aerial vehicle |
| UV | ultraviolet |

CHAPTER ONE

# Introduction

Travel and tourism is now the world's largest industry. American commercial air carriers alone generate over $100 billion in revenue annually and account for 720,000 jobs. Directly related industries, including commercial-aircraft production and airport services, generate an additional $50 billion and 375,000 jobs.[1] More broadly, air travel has become, for many Americans, an integral part of their professional and personal lives. In 2000, passengers on U.S.-owned commercial air carriers took more than 600 million trips.[2] Clearly, a credible threat to the viability of America's commercial airline industry could have profound effects on the nation's economy and on Americans' way of life.

Shoulder-fired missiles (also sometimes called man-portable air defense systems [MANPADS]), such as the U.S.-made Stinger, pose such a threat. These systems are relatively inexpensive, widely available in the international weapons marketplace, and lethal to aircraft lacking countermeasures. They offer terrorists a means of bringing down airliners at any airport and thus have the potential to induce air travelers to think that they are not safe anywhere.

This paper is intended to assist policymakers in formulating appropriate responses to the threat posed by MANPADS in the hands of terrorist groups. Specifically, we address the question of whether the U.S. government should develop and deploy countermeasure systems to protect commercial aircraft from such missiles and, if so, what types of systems merit consideration.[3]

The paper addresses the following topics:

- the nature and severity of the threat posed by terrorist groups with MANPADS and the potential economic consequences of missile attacks on U.S. commercial aircraft
- the MANPADS threat in the context of the broader war on terrorism
- operational and tactical aspects of the threat, including potential operating areas around major airports and options for limiting terrorists' access to weapons and potential targets

---

[1] Wilbur Smith Associates, "The Economic Impact of Civil Aviation on the U.S. Economy," www.wilbursmith.com/Economic_Impact_of_Civil_Aviation_on_the_US_Economy.pdf, accessed 5/1/03.

[2] That is, there were 600 million "revenue emplanements" on American air carriers in 2000. Air Transport Association, Office of Economics, Airline Industry Facts, Figures, and Analyses, http://www.airlines.org/econ/d.aspx?nid=1026 (as of March 1, 2003).

[3] At the time of writing, the Department of Homeland Security (DHS) is leading the Counter-MANPADS Air Defense System Development and Demonstration effort, aimed at migrating available military technologies that could best protect commercial airliners from MANPADS. The two-year effort began in October 2003 and is intended to result in recommendations to the administration and Congress on how to proceed. In addition, there were two congressional bills recently introduced related to the installation of countermeasures aboard commercial airliners (see Appendix B).

- the characteristics and limitations of technical options available for countering missiles once they have been launched
- the potential cost of developing, installing, and maintaining candidate countermeasure systems, along with a consideration of who should bear the costs
- recommendations regarding how to proceed

CHAPTER TWO

# The Threat: A Clear and Present Danger?

Just how serious is the threat posed by MANPADS in the hands of terrorists? Terrorist threats, by their very nature, are difficult to evaluate precisely. Enemy groups are constantly mutating, seeking to master novel capabilities, recruiting new foot soldiers, shifting locations, changing leaders, plotting different attacks. Under pressure from United States and allied security agencies, al Qaeda and related organizations may lose capacity in one dimension but gain in another. Factors such as these make it difficult to predict terrorist attacks with any specificity. However, it seems prudent for decisionmakers responsible for homeland security to regard the probability of an event as high when those who would perpetrate it have, at once, the motive, means, and opportunity to carry out the act. Do they?

**Motive.** This, at least, seems clear: terrorists from al Qaeda and its affiliates want to kill Americans, and they want to do so in spectacular ways.[1] Since the 1970s, international terrorists have exhibited a particular fascination with commercial aircraft, which are regarded as symbols of Western technological prowess. Bin Laden and his lieutenants must also recognize the strategic value of attacking Americans in their homeland. The U.S. response to al Qaeda's killings of Americans in Saudi Arabia, Africa, and Yemen prior to 9/11 was rather restrained. The reactions to 9/11 were anything but. Attacks on the American homeland have the potential not only to create larger numbers of U.S. casualties but also to yield much greater economic effects in this country than attacks overseas. Furthermore, occurring as they do under the very noses of U.S. security agencies, they can have profound effects on Americans' sense of security and well-being. In light of this, we judge that al Qaeda's leaders would relish the opportunity to bring down American commercial aircraft full of passengers, preferably in daylight and in cities that are major media hubs. Although they would expect to get the most value out of attacks perpetrated in the United States, they would also regard as useful the ability to attack American airliners abroad.

---

[1] While that is clear and sufficient for our purposes, it begs the larger question of *why* terrorists seek to kill Americans. Their own statements on this issue are not very enlightening. The leaders of al Qaeda and similar groups claim that they conduct such attacks in the name of Muslims everywhere in pursuit of objectives that have shifted somewhat over time. Prominent among these have been to force the United States to withdraw its military forces from the Middle East and to abandon its support of pro-Western governments there, to expel Israel from occupied territories, to overthrow secular governments in Muslim lands, and to reestablish the Caliphate. To those not in thrall to radical Muslim ideology, no meaningful connection between these millennial goals and specific terrorist attacks is discernible. Rather, at a practical level, the killing is undertaken both for its own sake (these people are professional murderers) and for reasons of institutional vitality. Terrorist organizations, like legitimate enterprises such as businesses, foundations, and universities, compete for money, talent, and influence. In the terrorist world, these goods tend to flow to groups that demonstrate frequently their potency as instruments for striking out against the enemy. See D. Benjamin and S. Simon, *The Age of Sacred Terror,* New York: Random House, 2002, pp. 156–66.

**Means.** Al Qaeda and many other groups hostile to the United States have MANPADS and the ability to use them. Over 700,000 MANPADS have been produced worldwide since the 1970s.[2] The United States and other countries provided MANPADS to mujahideen fighters in Afghanistan during the 1980s, along with hands-on training to ensure that they could be used effectively (which they were). Many thousands of MANPADS, including some Stingers sent to Afghanistan, are said to be unaccounted for worldwide. During the recent U.S. operations in Afghanistan, Russian SA-7s and British Blowpipes were recovered from Taliban caves in Afghanistan. SA-7s and other Russian-made models can be purchased in arms bazaars in a number of Middle Eastern and Central Asian countries. In some of these markets, such systems are sold for as little as $5,000.[3] With the increase in collaboration among terrorist groups, one may expect the transfer of a variety of MANPADS types among them. Figure 2.1 provides an overview of non-state groups known or thought to be in possession of MANPADS today.[4] Al Qaeda in particular has at least first-generation MANPADS, has the ability to move them about internationally, and has decided to employ MANPADS attacks as part of its terror campaign. That was shown, for example, by the November 2002 attempt to use two MANPADS missiles to bring down an Israeli charter airliner departing Mombasa, Kenya.

**Opportunity.** Herein lies the question. Creating opportunities for attacks means smuggling one or more missile systems and trained operators into either the United States or, failing that, a country regularly served by an American carrier, and positioning them for a high probability-of-kill shot against an arriving or departing flight. The fact that no known attempts have yet been made against a U.S. civil carrier suggests either that the required assets are not in place or that al Qaeda's leaders are waiting for what they regard as a more propitious time to undertake such attacks.

The difficulties associated with getting the assets in place are certainly not insurmountable for an organization such as al Qaeda. The difficulties and risks associated with smuggling a handful of man-portable weapons and a few trained operators into the United States (or neighboring countries regularly served by U.S. carriers) are probably commensurate with those of training, indoctrinating, and positioning the four teams of men who commandeered and flew the aircraft involved in the attacks of September 11. As their name suggests, these weapons are small and lightweight (less than 40 pounds). They could easily be smuggled into the United States in a packing crate inside one of the 20,000 uninspected shipping containers that are unloaded at U.S. ports every day or by a variety of other means.[5]

MANPADS have already been used by terrorists and other operatives against a variety of aircraft in many parts of the world, including Africa, Asia, and Central America. One source

---

[2] CSIS, "Transnational Threats Update," Vol. 1, No. 10, 2003.

[3] Jane's Terrorism and Insurgency Centre, "Proliferation of MANPADS and the Threat to Civil Aviation," August 13, 2003, http://www.janes.com/security/international_security/news/jtic/jtic030813_1_n.shtml (as of November1, 2003).

[4] See D. Kuhn, "Mombassa Attack Highlights Increasing MANPADS Threat," *Jane's Intelligence Review,* Vol. 15, No. 2, 2003, pp. 26–31. See also T. Gusinov, "Portable Missiles May Become Next Weapon of Choice for Terrorists," *Washington Diplomat,* June 16, 2004; and P. Caffera, "U.S. Jets Easy Target for Shoulder-Fired Missiles," *San Francisco Chronicle,* November 30, 2002, p. A14.

[5] An interesting example of an alleged MANPADS smuggling ring involves the ongoing case of the British national Hemant Lakhani. According to prosecutors, Lakhani agreed to deliver an SA-18 missile to U.S. agents posing as buyers after he obtained it from Russian agents posing as sellers. CNN, "Feds Tell How the Weapons Sting Was Played," CNN.com, August 14, 2003, http://cgi.cnn.com/2003/LAW/08/13/arms.sting.details/ (as of November 3, 2004).

**Figure 2.1**
**Proliferation of MANPADS among Selected Non-State Groups**

Different MANPADS classes available on black and grey markets

1st gen infrared – Reticle Scan, SA–7
2nd gen infrared – Conical scan SA–14, SA–16, Basic Stinger
3rd gen infrared – Pseudo imaging, SA–18
CG – Command guided, Blowpipe

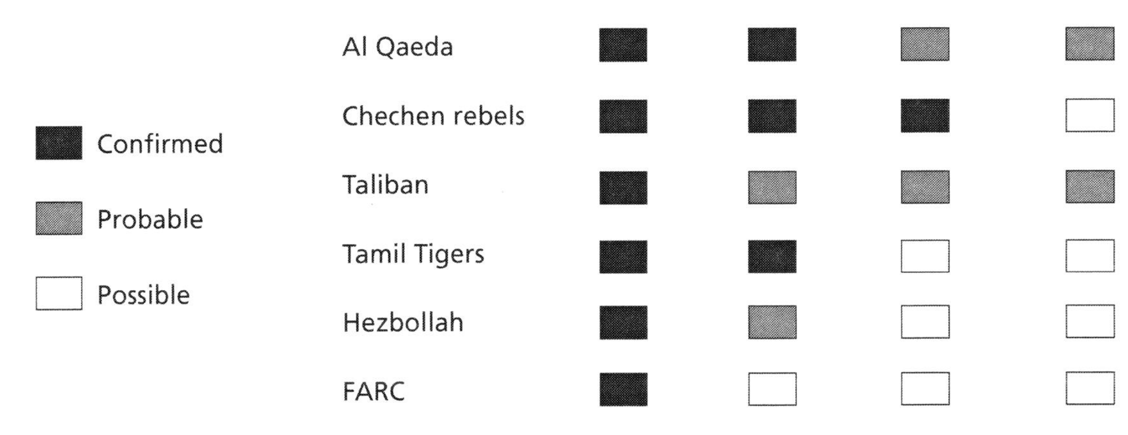


RAND *OP106-1*

estimates that of the 35 recorded attacks against civilian aircraft, 24 planes were shot down, killing over 500 people.[6] Most of these attacks, however, were against non-jet-powered aircraft, such as helicopters and turboprop and piston-engine aircraft. This same source lists only five incidents where large jet-powered airliners were believed to have been attacked by MANPADS, including the attack on the Israeli jet in Kenya. Of these, two of the five resulted in catastrophic losses. Most recently, in November 2003, a DHL Airlines Airbus 300 was damaged by a MANPADS while flying near Baghdad International Airport, but managed to return safely without loss. No attempts have been recorded against a U.S. commercial airliner.

Given the motive to attack commercial airliners with MANPADS, the means to do so, and the opportunity to bring weapons and operators into the United States, why have we not already witnessed attempted attacks on American commercial airliners? One answer seems to be that the terrorist leadership has thus far regarded other means of killing Americans as more attractive. Because commandeering an airliner and crashing it into a large building was feasible, this tactic was preferred because it would not only undermine Americans' confidence in flying but also would produce far more destruction on the ground. However, as measures are taken to preclude 9/11-style attacks (e.g., improvement in screening at airports, deployment of air marshals on aircraft, strengthening of cockpit doors), attacking aircraft with MANPADS will unavoidably become more attractive to terrorists.

---

[6] There are various estimates of these totals. The quoted numbers are taken from C. Bolkcom, B. Elias, and A. Feickert, *Congressional Research Service Report for Congress: Homeland Security: Protecting Airliners from Terrorist Missiles*, 2003.

CHAPTER THREE

# Potential Economic Welfare Impact from an Attack

Threats to commercial aviation are numerous and varied, and the cost of instituting preventive measures for all of these threats could become quite large. A sense of the economic impact of an attack affords some context for the allocation of resources to countermeasures. Economic losses may be divided into three categories: immediate, tangible losses from the attack; losses to travelers and airlines during a subsequent air-travel shutdown (as after the 9/11 attacks); and losses to travelers and airlines from reduced demand once the industry resumes operations.

Initial damages from such an attack would likely approach $1 billion per aircraft destroyed. These are straightforwardly estimated. Larger aircraft typically cost $200–250 million (depending on the exact model) and carry around 300 passengers each. Monetizing the value of the lives of the passengers aboard is always an uncomfortable calculation, and no earthly compensation can restore the loss of a loved one. But to make the tradeoffs that must be made in other situations between lives and resources, compensation policies and other economic treatments typically approximate a value per life of $2–2.5 million.[1]

In the aftermath of September 11, commercial air travel was stopped entirely for a few days and was severely disrupted for at least a week before flight schedules returned to something even close to normal. We infer that shutdowns of individual airports and the whole system are a possibility if a MANPADS attack were to be successful. In the remainder of this section, we show how we estimated shutdown losses—first for travelers and then for the airlines—and losses after resumption of operations.

## Estimating Shutdown Losses

Beyond the immediate destruction of life and property, the economic impact of a MANPADS attack can be characterized in different ways. One ramification is the potential change in gross domestic product (GDP) that may result from the airline shutdown itself, from slowdowns in industries associated with the airlines, and subsequently from changes in people's behavior. Lower demand for air travel may lead to less spending on other kinds of goods and services

[1] The September 11 Victim Compensation Fund made an average death claim payment of about $2.1 million. See Department of Justice, September 11th Victim Compensation Fund of 2001: Compensation for Deceased Victims, http://www.usdoj.gov/victimcompensation/payments_deceased.html (as of January 5, 2004). See also L. Dixon, *Assistance and Compensation for Individuals and Businesses after the September 11th Terrorist Attacks*, Santa Monica, Calif.: RAND Corporation, MG-264-ICJ, forthcoming.

in the economy, such as fewer stays at hotels and decreased business travel. Aircraft manufacturers (at home and abroad) may be adversely affected, too. However, while some activities may decline, others may increase. Domestic travel destinations may replace international ones. Individual cities or regions may be affected in different ways, but losses in some areas will be offset to some extent by gains in others.

We have chosen an alternative way to value the economic impact—through a measure of economic welfare. Economic welfare captures aspects of the value of air travel that may not be captured in GDP. Economic welfare is made up of consumer and producer surplus. The value consumers attach to the goods and services they buy is *at least* what they paid for them, or they would not have bought them. Consumer surplus is the excess of that value over the price. Analogously, producer surplus is the difference in value between what is paid to companies for a service (or good) and what it costs companies to produce that service. In this context, the welfare change is the change in the surplus value of transportation to air travelers and the change in profitability of the airline industry.

Consumer surplus may be understood intuitively as follows. Someone may pay $350 to fly round-trip across the country, but that person might have been willing to pay up to $500 for that service. The difference between what was paid for the service and how it was valued is the surplus the consumer gains from the trip—in our simple case, $150. The importance of consumer surplus is that if an attack happened and air travel were shut down, or if the airlines continue to fly but the traveler does not feel comfortable flying, the consumer in our example would be willing to pay $150 to be able to fly again with the system the way it was before. In that sense, the consumer values the loss of air travel (or his or her concern over its safety) at $150 for that trip.

To estimate the actual consumer surplus for a successful shoot-down of a commercial aircraft in the United States, we first divide air travel into different market segments based on length of trip. Then we estimate the cost of travel for different travel modes, including air travel and its alternatives, in these market segments. We examine travelers' willingness to pay to avoid the shutdown of air travel, which varies by segment, and calculate the consumer surplus loss that results. (For details of the consumer surplus estimate, see Appendix A.) We estimate that, in the event of a one-week airline shutdown, a consumer surplus loss of $2.0 billion would accrue; during a month-long shutdown, the loss would amount to $8.4 billion.

Producer surplus is defined as the difference between revenue and costs. Because passenger revenue should approximate passenger costs during normal operations in a competitive industry, we ignore lost profits in our calculations.[2] However, a systemwide shutdown would mean a number of costs with no revenue to offset them.

For short-term disruptions of a day or a week, we assume that contracts will obligate labor and capital costs (including leases on aircraft) to be paid. Only fuel costs would be saved,

---

[2] In economics, a competitive industry theoretically yields zero "excess" profit—that is, profit above and beyond a normal rate of return experienced across industries. This result holds because if these excess profits exist, new entrants will join a market and compete those excess profits away, until there is no greater incentive to join the industry in question than any other industry. A normal rate of return remains, but this just offsets the cost of capital (and usually compensates for the risk of the investment). Figures from the 2002 Annual Report of the Air Transport Association show that its member airlines earned a net profit of 1.3 percent of revenue over the period 1991–2000.

as planes would not be flown, but all other costs during this period would be lost. With a shutdown of a month, food expenditures might be avoided, but once again capital and labor and all other non-fuel costs would be counted as losses. With those assumptions, we estimate producer losses of $1.4 billion for a one-week shutdown and $5.6 billion for one month. The combined welfare loss (consumer and producer surpluses) would thus come to $3.4 billion for a one-week shutdown and $14.0 billion for one month.

## Losses after Resumption of Airline Service

Once airline operations resume following a shutdown, a significant amount of air travel may still be deterred through fear of flying, changes in airline operating schedules, and the increased inconvenience of additional security measures. That represents a loss to society in the form of further decreases in consumer and producer surplus. Reductions in future travel should be greater the longer the systemwide shutdown is. We assume that a shutdown of a day would reduce the number of flyers by 10 percent of normal in the following two-week period, with the corresponding loss.[3] A shutdown of a week would affect 15 percent of the travel for the next six months. And a one-month shutdown is taken to reduce travel for the next year and a half by 25 percent.

This 10–15 percent net reduction in travel corresponds roughly with the experience of the airline industry in the aftermath of the prohibition of travel for more than a day but less than a week after September 11. Airline layoffs announced through February 2002, six months after the system shutdown, were 14 percent of employment. And a year later, air travel was still down about 8 percent, although some of this decline is certainly due to a general slowdown in the economy.

These future losses from reduced demand for air travel can be quite large, larger in fact than the loss during the period of shutdown because the fear and uncertainty driving them lasts much longer than the shutdown. We estimate that the country would be willing to pay $12 billion to avoid an incident that would seriously affect travelers' confidence for the next six months. That value increases to over $50 billion for an incident that would affect travel for a year and a half. Admittedly, these future loss factors are somewhat speculative.

## Summary and Caveats

In summary, then, based on the effects of the attacks of September 11, we find it plausible that demand for air travel could fall by 15–25 percent for months after a successful MANPADS attack on a commercial airliner in the United States. A weeklong systemwide shutdown of air travel could generate welfare losses of $3–4 billion, and when losses from reduced air traffic in the following months are added in, the result could exceed $15 billion (see Table 3.1). By *losses*,

[3] This 10 percent is a net reduction in air travelers. Some travelers who were scheduled to fly during the shutdown will resume their intended travel after the shutdown ends, but the net reduction is 10 percent. Also the distance pattern of flights taken is assumed to remain stable—that is, no shift to shorter or longer routes.

**Table 3.1**
**Total Welfare Losses from a Systemwide Shutdown (in billions)**

| | One Day | One Week | One Month |
|---|---|---|---|
| Consumer Surplus Loss | $0.3 | $2.0 | $8.4 |
| Producer Surplus Loss | $0.2 | $1.4 | $5.7 |
| Direct Loss Subtotal | $0.5 | $3.4 | $14.1 |
| Future Loss Factor | 10% | 15% | 25% |
| Indirect Loss Subtotal | $0.9 | $12.4 | $56.6 |
| Total Loss | $1.4 | $15.8 | $70.7 |

we mean that airlines and the traveling public would be willing to pay that much to avoid a catastrophic attack on air transportation. Note that the amount is not MANPADS-specific; it represents the willingness to pay to avoid such an attack from any source of terrorist threat, be it MANPADS attack, hijacking, bombing, or other. If the public reluctance to fly were less severe or lasted for a shorter period of time following an attack (say, because of specific countermeasures to the threat that could be rapidly adopted), the welfare loss would be less. If the public reaction were even greater, the welfare loss could be even more.

A few other caveats to this economic analysis are in order. The analysis does not attempt to address any issues of local traffic congestion from changes in travel patterns. Nor does it account for the possibility that passengers may be willing to pay more per hour to avoid additional travel time as trips get longer; without specific data to rely upon, we presume these per-hour values to be uniform across all trip distances. Our estimates of consumer surplus loss depend on elasticity calculations—that is, estimates of the responsiveness of demand to price changes. These are best applied in a narrow range around the values for which they are estimated. The system shutdown scenarios produce effective price increases well outside of the typical range faced by business and leisure travelers. The use of the elasticity estimates here is done in the absence of either better data (or a more robust theory of traveler reaction to safety concerns) to inform the calculation, and this use is best confined to estimating an order of magnitude of results rather than attributing accuracy to specific numbers.

Finally, losses during a shutdown and following resumption of service are likely to be strongly conditioned by the success of law enforcement at apprehending MANPADS operators and their supporters. If arrests are made, federal officials can credibly assure the public that air travel is safe, and no further attacks follow the resumption of service, economic losses may be no greater than those shown here for a shutdown that might be as short as a week. If one or more of those conditions is not met, a longer or repeated shutdown and disproportionately larger post-resumption losses may accrue.

CHAPTER FOUR

# Strategic Considerations

Decisions regarding the installation of countermeasures aboard airliners must be considered within the context of the larger war on terrorism being waged by the United States and like-minded governments around the world. Al Qaeda and groups with similar capabilities and agendas constitute a serious threat not only to the safety and well-being of Americans but also to America's position in the world. Global-reach terrorism and the battle against it is seen as a contest played out on several fields at once in which the audience is global and the primary stakes are psychological. Both sides try to shape the perceptions of this global audience: terrorists seek to convince people that their cause is just and worthy of support; responsible governments seek to spread the conviction that terrorist attacks are immoral and that they run counter to the interests of the terrorists' potential supporters.

Seen in this light, it is clear that one or more successful attacks on American commercial aircraft would have profound strategic consequences for the United States and its partners in the fight against terrorist groups. America's enemies would gain a tremendous psychological boost from such attacks and would confront the world's population with serious doubts about not only the safety of air travel but also the viability of their governments' counterterrorism efforts. A new front would be opened in the contest and the effects would be long-lasting: in the popular imagination, the terrorists would be credited with having the capability to kill people on commercial aircraft more or less at will until such time as convincing policy solutions to the threat were implemented.

Of course, no countermeasure (or combination of countermeasures) can reduce to zero the possibility that terrorists could bring down an airliner with MANPADS. Less than perfect countermeasures have proven highly valuable in the context of military operations, where each aircraft is expected to encounter enemy defenses many times during its operational life span and where some risk of loss is accepted. The choice is less clear-cut in this case because the probability of an attack remains, by comparison, quite low, and because it is not clear whether the installation of less than perfect countermeasures will be sufficient to convince the vast majority of the public to return to flying.

The solutions we consider in this paper might render a MANPADS attack futile or even counterproductive. However, because of the complexity of technical countermeasures, it may take years before those solutions can be effectively designed and implemented. While that leaves open a window of vulnerability, waiting to start development until after an attack occurs would leave the defense of air travel that much farther behind. Thus, as we argue below, steps to develop them should proceed quickly.

CHAPTER FIVE

# Policy Solutions and Operational Issues

While this paper focuses on the question of countermeasures installation, there are other potential policy responses to the MANPADS threat that deserve mention. These responses are not mutually exclusive, but rather could act in concert to create a layered defense. As shown in Figure 5.1, the layers begin at the most distant remove by affecting the supply of terrorists and weapons (shown at the bottom of the figure) and range inward with increasing criticality to those that address post-incident crisis response (working to the top of the figure).

No individual solution will completely remedy the problem. Addressing it on a number of levels could decrease a potential attacker's confidence in the utility of MANPADS. The hope is that enough uncertainty about success will dissuade terrorists from choosing MANPADS as an attack means.

Taking the war to the terrorists' homeland seizes the initiative.[1] Offensive operations taken by the United States against terrorists where they are based (e.g., against al Qaeda in Afghanistan during Operation Enduring Freedom) is an important example. Striking and capturing terrorists not only affects the MANPADS threat but also other parts of the terrorist system. More focused options include buyback programs and technology-control regimes directed at MANPADS, which should help reduce current and future threat potential. To the extent that such initiatives do not keep MANPADS out of terrorist hands, there will be a need to improve security along borders and at transportation hubs to interdict the movement of people and weapons. Assuming these were not enough and that terrorists were still able to place MANPADS in the United States, we get to the final four layers of the solution space: preventing MANPADS from being fired, preventing a launched missile from hitting the aircraft, minimizing the damage from a missile hit, and minimizing consequences from an attack.

To prevent MANPADS from being fired, one could try to secure a perimeter around an airport that would prevent an attacker from firing from within range of the missile system. The range of a system like the SA-7 can extend out to 3.5 miles and a maximum altitude of 10,000 feet.[2] So where would the security perimeter need to extend to prevent a launch? To get a better sense of this, we need to consider the routes airliners fly when landing or taking off from an airport.

---

[1] We restrict ourselves here to security and military operations. We do not mean to slight the importance of fighting terrorism by addressing through political and economic means the conditions giving rise to it.

[2] Tony Cullen and Christopher F. Foss, eds., *Jane's Air Defense Systems, 2001–2002,* Surrey England: Jane's Information Group, 2001

**Figure 5.1**
**Protection Against MANPADS Provided at Many Levels**

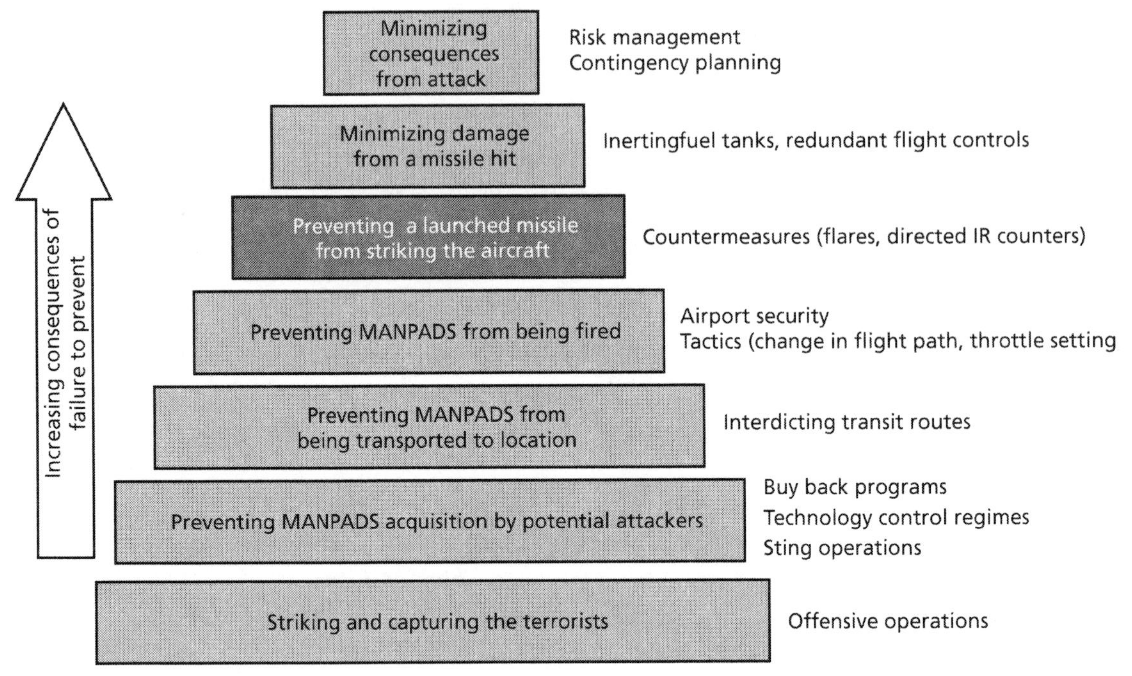


RAND *OP106-2.1*

We were able to obtain, through public sources, standard arrival and departure patterns for Los Angeles International Airport (LAX). Noise regulations and other ordinances make these patterns available for many airports across the country. These patterns describe where and how low airliners fly in the airport's vicinity. With this information, we were able to define the area within which a terrorist armed with an SA-7 could pose a threat to an airliner. These findings are generally and broadly applicable to urban airports. We found that terrorists using that kind of MANPADS may engage aircraft while situated anywhere within an 870-square-mile area of the Los Angeles region. A more modern MANPADS (for example, the SA-18) has the capability to engage a slow-flying commercial aircraft up to 18,000 feet, which would allow the terrorist to be located anywhere within a 4,600-square-mile region. Against either the older or more modern threat, *completely* preventing an attack solely through the use of enhanced security perimeters would be impractical, considering the large urbanized areas involved, the cover provided by urban structures, and the availability of multiple freeways for quick access to attack and getaway (some of the flight paths extend over Santa Monica Bay, where a terrorist could engage an airplane from a small boat). However, since the probability that most MANPADS will hit a target drops rapidly when fired near their maximum range, the security emphasis might be placed on preventing launches from closer ranges—for example, near the airport. Secure perimeters close in could impede shorter-ranged threats of various types. As an example, a possible threat reaction to the installation of MANPADS countermeasures might be to use simpler weapons that are not affected by countermeasures, such as small arms or rocket-propelled grenades (RPGs) fired from a parking lot 100 feet under a runway approach.

In short, airport perimeter security is insufficient as a stand-alone defense against MANPADS but could serve as one of a number of layers in an overall suite of protection measures. In this context, it has the potential to blend in nicely with countermeasure-based solutions. Airport perimeter security also offers protection against other threats (e.g., the possibility of attacks on the airport itself).

If the layer attempting to prevent a missile launch fails, countermeasures might prevent the missile from making a successful hit. The basic types of countermeasures are discussed in more detail in the next section, but broadly speaking, different systems each have their own pros and cons. Some systems could provide highly effective protection under a wide range of conditions, but none are able to protect against the full range of threats. An understanding of where countermeasure weaknesses lie should help focus other counter-MANPADS policy efforts. As an example, if the countermeasures deployed are known to be marginally effective against a class of threat systems, buyback programs and nonproliferation efforts should pay particular attention to those systems. If we have an understanding of which types of aircraft are most vulnerable to MANPADS, this can help inform which aircraft should be fitted with countermeasures (or which should be fitted first).

By aircraft *vulnerability* to MANPADS, we mean the level of aircraft damage resulting when a missile strikes. Minimizing that damage is the next layer of defense. Vulnerability is an important consideration about which little is known in the case of airliners. There has been some renewed interest in the area, but at very low levels of funding.[3] A good deal of research and testing has been done on the vulnerability of military transport aircraft, but these aircraft can be significantly different in design from commercial airliners: in many respects, military transports are designed to reduce their vulnerability to weapon fire. We do know that modern commercial airliners are designed and aircrews are trained to fly with one engine inoperable. MANPADS with infrared (IR) seekers are drawn to hot emissions or parts, such as those found on or near jet engines, so given the paucity of actual data, one might suppose that a MANPADS hit would at least disable an engine.

The aforementioned attack on the DHL aircraft departing from Baghdad International is instructive in this regard. Amateur video shows the missile type and launch geometry. This example suggests that the effects of a MANPADS hit can be more complex than the loss of an engine: in this case, it was reported that both engines of the aircraft were operable, but that all flight hydraulics were lost, primarily from fire-induced damage. There are technologies available that can help limit damage from fire, such as gas generator systems, which remove the highly flammable vapors in a fuel tank and replace them with a nonflammable inert gas. It would be expensive to retrofit these systems into existing airliners, however, and despite the DHL experience, the likelihood of extensive fire damage from a MANPADS hit is still unknown. Finally, as discussed above, the magnitude of the indirect losses from a successful MANPADS attack will depend on the ability of the authorities to reestablish the security of air travel and quickly convince travelers that they have done so. Contingency planning across law enforcement agencies could increase the probability that perpetrators are captured or killed.

[3] See C. Pedriani, "JASPO/NASA Cooperate to Improve Commercial Aviation Security," in *Aircraft Survivability: Reclaiming the Low Altitude Battlespace*, Arlington, VA: Joint Aircraft Survivability Program Office, 2003.

Agreements may be reached in advance to implement various perimeter security measures that might be viewed as too costly or intrusive in the absence of an attack.

Planning and risk-management activities by officials not directly involved in law enforcement could also be helpful. For example, arrangements may be made in advance to alter aircraft approach and takeoff patterns in the event of a MANPADS attack. It will be essential that the messages the American public will be hearing from top homeland-security officials be consistent and accurately reflect the best knowledge available about risks. The large potential indirect losses we project are entirely due to actions taken out of perceived fear of attack. These losses can be reduced to the extent the fears are allayed. But if fears are falsely allayed—that is, if other attacks ensue following government assurances—the credibility of the government and its ability to manage risks could be severely damaged.

CHAPTER SIX

# Countermeasure Systems

We here consider three major categories of countermeasures to MANPADS that are either deployed or under development: flares, laser jammers, and high-energy lasers (HELs). The first two aim to confuse the IR seeker of an infrared missile, while the HEL aims to destroy the missile, regardless of how it is guided. In this section, we discuss each in terms of basic operation, effectiveness, robustness vis-à-vis counter-countermeasures, and sensor support requirements. There are other potential countermeasures that will not be discussed, including defensive missiles (airborne or ground-based) and airborne lasers with sufficient power to destroy the seeker head. These will not be available in the near- to mid-term, and in any event do not appear to be well suited to civilian applications.

## Flares

We describe three kinds of flares: conventional, advanced, and covert. *Conventional* flares were initially fielded to counter first- and second-generation passive IR missiles employing so-called seeker reticles: spokelike masks that rotate in the seeker's optical field of view and permit homing on the target. Conventional flares are intended to produce an IR signature so large that the target signature is overwhelmed, and the seeker locks onto the flare instead of the target. In quantitative terms, one speaks of achieving a high *jammer-to-target ratio* in order to capture the seeker.

Flares may be released either preemptively (before the onset of an attack) or reactively, after an IR surface-to-air missile (SAM) launch is detected. In a military setting, knowledge of when an aircraft enters a combat situation or arena can minimize the length of time that preemptive flares need to be released for. In the case of terrorists, such knowledge can be difficult to predict, and so for commercial applications, reactive flares are the practical consideration. In deployed systems, launch detection is usually accomplished by an optical or radar sensor onboard the aircraft; however, the cost-effectiveness of ground basing is receiving some scrutiny in recent studies. One drawback of ground-basing is the requirement for a highly reliable communications link from the sensor to the aircraft.

Seekers on some second- and third-generation IR SAMs are able to discriminate flares from aircraft due to the free-fall flight profile of the flare or its spatial extent, spectral properties, or intensity profile. As an example, modern two-color seekers can discriminate conventional flares from airplanes based on their spectral signatures (i.e., the relative signal strength in different wavebands, or colors). The ratios of intensities across different bands are indicative of temperature, and flares are generally hotter than aircraft engines. *Advanced* flares can counter this discriminant because they consist of an ensemble (cocktail) of flares, each peaking in a different waveband, such that the combined signature matches that of the aircraft. Research is underway to replace cocktails with new single materials that can match target spectral signatures.

Some modern pseudoimaging seekers are able to discriminate against point targets such as flares by computing their aimpoint from the weighted centroid of the presented signature, including that of the airplane.[1] This discriminant can be negated by continuously dispensing advanced IR chaff, fabricated from pyrophoric materials, which react with atmospheric oxygen to release heat. Continuous dispensing can generate a chaff trail with sufficient extent to offset the centroid away from the target. The pyrophoric reaction is not sufficiently exothermic to render the material incandescent, so it is thought less likely that low-altitude release will cause fires on the ground than similar release of other flares. There is also no appreciable visible signature, leading to the common designation *covert.* Because of their covertness and reputed safety from fires, they are viewed as more suitable than other flares for installation on a commercial aircraft. The effectiveness of covert flares used in a *reactive* mode requires further development, however.

Flares, whether conventional or advanced, have little prospect for countering imaging seekers, which may be fielded by technologically advanced nations between 2005 and 2010. Flares are also ineffective against existing laser beam riders, which home in on a laser spot placed on the target by the SAM operator. SAMs that are radio-frequency (RF) command-guided (CG), like Blowpipe, are also largely immune to flares. The operation of such CG missiles is somewhat harder to employ effectively, since it requires users to keep the missile on an optical track between themselves and the target.

Since conventional flares could cause ground fires if released below about 1,000 feet, missile-warning system (MWS) used in conjunction with flares must generate few false alarms. Ultraviolet (UV) sensors, which are prone to false alarms, are thus not good candidates. Fusing multiple, independent phenomenologies (e.g., IR and Doppler radar sensors) have been proposed as a means to achieve an acceptable false-alarm rate. A Doppler radar measures the missile's radial velocity (i.e., its speed in the direction of the sensor). When viewed from the target aircraft, the missile's radial velocity shortly after launch is an unambiguous discriminant.

However, if a Doppler radar is deployed on the ground, it may face some delay before discriminating between SAMs and ground vehicles on a neighboring highway. Sensors looking normal to the missile's trajectory plane will initially measure zero Doppler, and seconds may be lost before the radial velocity exceeds the upper limit for vehicular traffic. This is particularly of concern if the aircraft is attacked at low altitude, which offers little time to respond. Employing steeper takeoff and landing profiles could be used to shrink the region susceptible to low altitude attacks, but the impact on safety of this option has not been fully evaluated. Providing geometric diversity by increasing the number of ground radars can ameliorate the problem, though at increased cost.

---

[1] Pseudoimaging seekers can coarsely resolve light sources within the field of view (FOV), either by scanning the FOV with a detector having a relatively small instantaneous FOV, or by employing a focal plane array (FPA) with a small number of detector elements. Some pseudoimagers also use two-color detectors to provide additional flare-rejection capability.

Key advantages of flares are that they are available today, they are fairly robust against even large salvos of older IR SAMs (which are the most highly proliferated), and they can be deployed based on detection of missile launch alone, not requiring sensors for tracking.

## Laser Jammers

Laser jammers, which will soon be commercially available, are the most advanced form of directed infrared countermeasure (IRCM), or directed infrared countermeasure (DIRCM).[2] They will work best against first- and second-generation MANPADS. Their objectives are first to overwhelm the signal produced in the enemy missile's seeker by the target, and then to substitute a specially modulated signal[3] transmitted by the laser, so as to divert the missile. The laser signal must emit at the color the seeker expects to see, be pointed with sufficient accuracy to enter the seeker optic, and achieve a large jammer-to-target ratio. Since the catalog of threats includes a variety of potential colors, a multiband laser or group of lasers is required for full protection. Laser spectra can be very narrow, but it is preferred that the DIRCM laser have a relatively broad spectrum to defeat narrowband optical filters that could be inserted in the seeker optic to block a jammer. Some DIRCMs employ optically pumped oscillators to jam the threat bands, and this technique typically results in broader spectra. The laser modulation is designed to capture the seeker of a MANPADS and to break the seeker's lock on the targeted aircraft.

Laser-jammer systems are complex due to the need for highly accurate pointing. Satisfying this requirement demands that tracking sensors be mounted onboard the aircraft. Current practice is to also perform initial detection from the aircraft, although there are some potential advantages for locating this function on the ground. Following initial detection by an MWS, a fine tracking beam (e.g., a laser radar) is slewed towards the SAM, performs a limited search to acquire the missile, and then maintains a close track while the modulated laser illuminates the seeker. First-generation systems will employ turrets to slew the tracking and modulated beams. Eventually, this function may be performed by laser arrays, or micro electro-mechanical systems (MEMS) based optical elements. In the meantime, the turret is likely to be the most failure-prone component in the system.

Stringent requirements for the MWS are high probability of detection and high accuracy. The tendency of the MWS to generate false alarms must also be taken into account, for three reasons. The first is that false alarms could lead to laser illumination of objects other than MANPADS and thus possibly to blinding of observers on the ground (this particular example would also require a false positive from the fine tracker). Laser eye-safe ranges for the DIRCMs

[2] DIRCMs employing high-intensity lamps as sources have been deployed on aircraft in the past. They are inferior to laser jammers in several respects. The lamp is an incoherent light source and cannot deliver the small spot size, high intensity, narrow spectrum, and modulation flexibility of a coherent laser source. These deficiencies are elevated in importance when the aircraft signature is large, as with large commercial aircraft. Lamp-based jammers provide inadequate jammer-to-target ratio to confidently protect large jet-powered commercial airliners.

[3] That is, the signal's variation over time in amplitude and frequency has been specifically designed to maximize its potential to confuse the enemy missile's seeker.

being tested are on the order of several hundred feet. Because this is less than the minimum range of IR SAMs, the MWS could be set to ignore objects at such close range, which should rule out damage to the unaided eyes of persons on the ground. Evidence we have received to date concerning observers using binoculars appears contradictory. It should be noted that the Air Force has not experienced any eye-safety incidents during employment of militarized versions of these systems and does not limit the operation of these systems in any portion of the flight regime for eye-safety reasons. The second reason is that false alarms lead to slewing of the turret, which may ultimately shorten the time-to-maintenance or time-to-failure of this key component. The final reason is that false alarms could set off the contingency plans of local law enforcement, airport authorities, and airliners, which will limit their effectiveness during actual firings and accumulate in cost over time. There is no way to mitigate this problem, so minimization of false alarms will be an objective in MWS design.

Optical MWS sensors fielded thus far have typically operated in the UV "solar blind" region of the spectrum. This is the UV band in which upper atmospheric ozone almost completely absorbs solar radiation. In the absence of a missile plume, the sensor can be triggered by only a few manmade and natural sources, including high-intensity lamps, aircraft afterburners, corona discharges, and lightning. Unfortunately, these sources are not rare in urban areas.

Proposals for improving the MWS false alarm rates have included emplacing the UV sensors on the ground (false alarms looking skyward are presumably lower); adding a different phenomenology detector, such as a Doppler radar or one-color mid-wave IR (MWIR) sensor; or replacing the UV detector with a two-color MWIR detector. MWIR sensors employing large focal plane array (FPA) detectors are on the verge of supplanting UV MWS systems on the next generation of fighter aircraft, though MWIR false alarm rates remain a contentious subject. Manmade sources of MWIR false alarms are more numerous than UV sources, but the high resolution of FPAs may enable the MWIR sensors to kinematically discriminate the stationary sources.[4] An important advantage of MWIR is its immunity to absorption by ozone in the lower atmosphere, which can be problematic for UV sensors in the urban environment.

In addtion to fasle alarm rate issues, the sequence of events following initial detection by the MWS, which includes slewing of the turret, fine tracking, and then a dwell period to break the seeker's lock, requires that the turret focus on one threat at a time. The DIRCM has some limitations against multiple threats, and though one could equip an aircraft with multiple turrets to increase the number of near-coincident launches that can be defended, this would obviously increase installation and operating costs by nearly that multiple.

As with flares, laser DIRCMs are not effective against laser beam riders (for which they may only furnish a beacon), RF CG missiles, and future imaging IR seekers. Current research is exploring whether IR focal planes might be disabled or degraded with increased laser power, but this is speculative and represents a departure from the basic DIRCM concept.

---

[4] That is, use the motion of different objects relative to the airborne MWIR to determine which are actually stationary on the ground.

To sum, a single-turreted laser-based countermeasure system would have good effectiveness against single shots by the majority of current MANPADS threat types and some dual coordinated firings but would not fully protect against all possible attacks.

## High-Energy Lasers

It was recently reported in the press that Northrop Grumman's ground-based mobile tactical high-energy laser (MTHEL) test-bed has destroyed artillery shells and Katyusha rockets in flight. The rocket is almost certainly a more hardened target than a SAM, which suggests that high-energy lasers might be used to protect commercial aircraft from shoulder-fired missiles in the vicinity of airports.

A palletized variant[5] of MTHEL, called Hornet, has been proposed for a wholly ground-based defense against MANPADS. The Hornet system would include a radar air picture to designate vectors along which the laser could not fire because friendly air traffic might be in the line of sight; netted IR search-and-track (IRST) systems for acquiring and tracking SAMs,[6] and for pointing the laser; and a megawatt-class deuterium fluoride chemical laser weapon housed in a turret on the ground.

Advertised performance of a single Hornet site indicates capability to defend against salvos of three missiles out to a range of at least five kilometers, with single-missile protection out to ten kilometers.[7] Robust protection of a large airport such as Reagan National would require a minimum of three sites. This assumes flight-corridor adjustments to keep aircraft above the SAM ceiling except when required for landing and takeoff. Many more would be required without corridor adjustments. It is important to consider that aircraft-based countermeasures do not typically invoke the need to include flight corridor adjustments.

The primary advantages of HELs for SAM defense are the ability to counter every current and future seeker technology, the robustness of a lethal kill as compared to smart jamming of the seeker, and the potential for defending against a wide variety of threats, including artillery, rockets, some cruise missiles, and hostile unmanned aerial vehicles (UAVs). HELs cannot operate under all weather conditions. Conditions that render the HEL inoperative will usually deny capability to MANPADS as well, but this is not true all the time. A spatially inhomogeneous fog layer may occasionally shut down the laser while leaving an open patch in which the SAMs can launch. At such times, unprotected flight corridors would have to be closed if no window of opportunity is to be left for MANPADS attack.

Eye safety is a concern when firing high-powered lasers, even if the lasers operate in the eye-safe band, as do deuterium fluoride lasers. Eye damage could arise either due to direct illumination of a person in an aircraft, or secondarily when persons in aircraft or on the ground see the

---

[5] That is, one packaged for installation aboard a vehicle.

[6] Including initial detection sensors onboard the aircraft would not be incompatible with the Hornet concept and might improve response time in built-up areas.

[7] Fewer missiles can be destroyed at greater distances because the dwell times have to be longer to compensate for the dissipation of energy.

destruction of the missile. The former problem should largely be circumvented by using the radar air picture to set up keep-out zones for the laser. There may be a rare occasion when untracked aircraft are illuminated, because air-traffic control radars do not detect with absolute certainty. However, since the laser beam must be slewed rapidly to keep pointed on the missile, any chance illuminations would last only milliseconds and would not cause damage. Diffuse reflections coming off the missile while it is illuminated can reach damage-level intensity only at a range of a few meters, which is clearly avoidable.

Perhaps the major drawback for HEL technology is availability. Estimates are that the start of production is at least three years away. This compares with current or imminent production for advanced flares, and for DIRCMs, though the latter's production *rate* may still be at issue.

Another concern with all the countermeasure systems discussed involves technology sharing and classification issues. Laser jam codes, sensor processing algorithms, and HEL systems are all sensitive technologies, which would need to be guarded. Assuming U.S. airliners would need to be protected during overseas flights (which many argue are the most vulnerable to terrorist attack), the question is how to maintain the systems and guard their classified information while in a foreign commercial-airport environment. This would particularly be troublesome for ground-based defenses such as the HEL system.

Figure 6.1 summarizes the effectiveness of flares, laser jammers, and HELs against different threat types.

**Figure 6.1**
**Summary of Potential Counters to MANPADS**

| Threat type | Proliferation | Countermeasures | | |
|---|---|---|---|---|
| | | Flares | Laser | High Power Laser |
| Older generation infrared (IR) |  Very wide | Demonstrated | Demonstrated | Potential |
| Current generation IR |  Wide | Potential | Demonstrated | Potential |
| Radio Control |  Limited | No Effectiveness | No Effectiveness | Potential |
| Laser Beam Rider |  Limited | No Effectiveness | No Effectiveness | Potential |
| Future IR (imagers) | None | Limited | Limited | Potential |

● Demonstrated ◐ Limited ○ No Effectiveness ◌ Potential

RAND *OP106-2.2*

CHAPTER SEVEN

# Costs

In this section, we estimate the cost of one class of MANPADS countermeasure described earlier: the laser-based DIRCM. Flares have the potential to be a cheaper alternative to laser jammers, but their effectiveness in a commercial-airliner application has more question marks. Ground-based HEL systems have excellent potential against current and future threats, including non-IR-based threats, but are not as far along as the laser-based DIRCM. It follows that remaining development costs in a commercial-airliner application would be more lengthy and expensive, although the overall *system* benefits may outweigh these initial costs. Even for aircraft with large IR signatures, laser jammers promise substantial capability against first- and second-generation MANPADS systems, due to the high signal-to-target ratio provided by the focused energy of the laser. From a technical-maturity standpoint, significant testing and development of this class of system have been done for the military (including live-fire tests in realistic conditions). It is therefore a front-runner for consideration of any kind of fast-paced near-term countermeasures installation program for commercial airliners. Though we restrict our specific cost estimates to DIRCMs, some of the cost issues highlighted in this section are relevant to the installation of other countermeasure types.

Total life-cycle cost (LCC) estimates can be broken down into two categories: installation costs and operating and support (O&S) costs. The LCC estimates are summarized in Table 7.1 for a projected quantity of 6,800 U.S. commercial aircraft. We examine estimates for these two cost categories in further detail below, and then conclude by considering total program costs in the context of the federal counterterrorism budget.

## Installation Costs

We made "first cut" cost estimates for installing the most promising near-term airborne DIRCM systems on the fiscal year (FY) 2003 inventory of approximately 6,800 U.S. commercial aircraft.[1] These costs were based on modifying the most current set of parametric cost data values from technically analogous military systems, such as the Air Mobility Command's large-aircraft IRCM (LAIRCM) system.[2] Specifically, we adjusted weights and volumes calculated

[1] The FY 2003 U.S. Aviation Inventory forecast was extracted from B. Turner, "FAA Aersospace Forecasts, Fiscal Years 2003–2014," February 13, 2003, Table I-2.

[2] RDT&E and procurement budget data on two Air Force programs, the laser-based LAIRCM system and the Special Operations Force's (SOF) AN/AAQ24 (V) 6 (lamp-based) DIRCM system, were extracted from the "FY-2004/2005 President's

**Table 7.1**
**Total Airborne DIRCM System LCC Estimates (FY 2003 dollars, billions)**

| Cost Element | Estimate (FY-2003 B$) |
|---|---|
| Installation | $11.2 |
| • RDT&E | $0.45 |
| • Production Start-Up | $0.17 |
| • Initial Spares and Test Benches | $0.90 |
| • A & B-Kit Procurement & Aircraft Retrofit (Based on Qty of 6,800) | $9.75 |
| • O&S (Phase-In and Ten-year Service Life After FOC)[a] | $27.0 |
| • A & B-Kit Maintenance | $12.5 |
| • Added Fuel | $4.2 |
| • Cost Growth/Uncertainty (25 Percent) | $4.2 |
| • Tech Upgrade Sustainment Cost | $4.1 |
| • Net Revenue Loss of Delayed Passengers | $2.0 |
| Total LCC Estimate | $38.2 |

[a] If an RDT&E phase begins in FY 2004, the first year of procuring DIRCM-modification kits for retrofitting commercial aircraft is assumed to begin in the FY 2007 time frame. Phase-in of O&S costs for the first configured aircraft begins in this fiscal year and continues until the last commercial aircraft is retrofitted in FY 2013. O&S cost continues once full operational capability (FOC) of all aircraft is completed in FY 2014, and costs are estimated annually for a ten-year service life through FY 2023.

for the military version to meet the form and fit required to enclose all the electronics within a canoe-shaped pod installed on the underside of commercial aircraft.[3] The airborne DIRCM systems proposed for commercial aircraft comprise

- an MWS of four two-color MWIR sensors capable of detecting a MANPADS approaching within the full range of velocities and angles possible
- a system processor designed to military specifications using fifth-generation or better central processing unit electronics along with a smart jamming card
- an electronic control unit that conditions the power and signals for the laser transmitter
- a multiband laser transmitter mounted within a small turret
- a command, control, and communications system to provide missile warning updates and intercept data to ground control operations
- a built-in-test hardware and software subsystem
- the canoe-type surface mounting hardware package and other A-kit interface hardware to enclose the preceding system components[4]

We estimate a total fleet installation cost of $11.2 billion.[5] That includes the

Budget Item Justification" sheets, February 2003. In addition, installation plans were outlined for the latter SOF DIRCM system as part of J. Townsend, "15 SOS Field Support Visit Aircraft Modernization," unclassified briefing, HQ AFSOC/XPQA, Halburt AFB, January 23, 2001.

[3] The parametric cost data and procurement estimates were based on the Navy's tactical aircraft DIRCM system data that was part of M. Popp, "Cost Analysis Update," briefing to the Interagency Task Force, NAVAIR 4.2V, Washington, D.C. February 13, 2003.

[4] A-kits are defined as the aircraft installation equipment used to attach and complete any wiring of the countermeasures to an airframe. B-kits are defined as the actual countermeasures equipment without installation.

[5] All costs in this section are given in FY 2003 dollars.

- research, development, test, and evaluation (RDT&E) phase at $445 million, consisting of the systems design, aircraft flight testing, and Federal Aviation Administration (FAA) certification[6] for six fully configured prototype systems
- manufacturing technology, capital, and facilities costs at $165 million to build up the annual production rate needed through the end of low-rate initial production (LRIP) ($65 million), and the set-up tooling of a second final assembly and test manufacturing line for the entire full-rate production (FRP) phase ($100 million)
- purchase of initial spares and test bench equipment at $900 million
- procurement and retrofit of approximately 6,800 DIRCM B-kits and A-kits at $9.75 billion

We assumed the development phase of a commercial DIRCM system begins in FY 2004 with the first year of RDT&E annual funding for the program office and continues for four years, until the end of FY 2007. The development estimate assumed the six flight-test prototypes would be adequate for integrating and checking out all the structural, electrical, and other interfaces required across the most representative subset of commercial aircraft models. In addition, the estimate includes a sufficient number of prototype ground and flight tests to ensure that the acoustic, vibration, and other environmental conditions of each aircraft model are within acceptable limits for the system to operate effectively. Finally, the total RDT&E estimate includes an adequate number of reliability and maintainability demonstrations within commercially acceptable threshold and objective values, especially for the on-equipment maintenance turnaround times across each of the different commercial aircraft models (as explained below, these could affect O&S costs if the targets are not met). To allow for these activities, we estimated the commercial DIRCM system development cost by increasing the LAIRCM RDT&E total budget by 60 percent (i.e., another 1.6 times the LAIRCM budget will have to be spent to fund DIRCM RDT&E for commercial aircraft). That factor was based on the following considerations:

- extent of repackaging of the B-kits to fit within the canoe structure
- number of unique A-kit designs needed for each commercial aircraft model
- number of flight tests required for installing, testing, and certifying the systems on all the commercial aircraft models

Procurement is projected to begin with an LRIP phase starting in FY 2006 during the third year of the system development phase and continuing through FY 2009. By then, approximately 1,100 commercial aircraft would be fitted, enough to cover all three stages of civil reserve airfleet (CRAF) deployment,[7] as well as all long-haul large jets for international and domestic flights. FRP will cover the remainder of the civil aviation fleet and will start immediately after the end of LRIP in FY 2009 and proceed through FY 2013. The cumulative average unit

[6] The FAA would be required to issue Supplemental Type Certificates that have demonstrated that DIRCM systems are capable of operating without conflicts or problems.

[7] Even though only a handful of designated commercial aircraft may have been deployed to support the last several conflicts in Iraq and Afghanistan, the U.S. Transportation Command, with approval of the Secretary of Defense, has authorized quantities of commercial aircraft that can be activated for all three stages of CRAF. The quantity of CRAF aircraft over the three stages is listed as part of Air Transport Association, Office of Economics, June 21, 2003, http://www.airlines.org/econ/p.aspx?nid=6342 (as of November 3, 2003).

production cost (AUPC) is estimated at $1.3 million (in FY 2003 dollars). This cumulative AUPC is the sum of the unit cost of the following:

- airborne DIRCM system A-kit, B-kit, and system installation cost
- the cost of the initial spares, technical data, support equipment, and change orders, amortized on a per-system basis by applying a 92 percent cost improvement curve or learning curve slope[8] across the total quantity of approximately 6,800 systems.

## Operating and Support Costs

We estimated an annual O&S cost per aircraft of $300,000 for a subtotal O&S cost of $27 billion through FY 2023. These costs consisted of the following:

- added fuel cost (of $45,000 per aircraft per year) needed due to the drag and additional weight that the commercial aircraft will be carrying over an assumed 3,000 hours per year
- maintenance cost (of $140,000 per aircraft per year) of the
  - airline mechanics labor, spares (following the initial buy), and other material needed to do on-equipment airport ground maintenance
  - airline depot-level or contractor logistics support activities for scheduled system overhauls and repairs and for unscheduled repairs of failed components sent back from the airport
- technology upgrade sustainment costs (of $60,000 per aircraft per year) that maintain the capability of the countermeasures as different threats emerge (however, this would not account for dramatic shifts in threat capability)
- operations costs due to airplane delays (of $15,000 per aircraft per year)

The added fuel cost estimate is based on an increased drag estimated at 0.4 percent and an added system-level total weight (including a 25 percent margin) of approximately 500 lbs. divided between the following:

- B-kit estimated weight of 125 lbs.
- weight of the canoe and other A-kit hardware (airframe structural material, wiring, doublers, isolators, etc.) estimated at 375 lbs.[9]

The maintenance cost estimates were driven by

- a mean-time-between-failure (MTBF) estimate of 800 hours, based on projected military-system reliability

---

[8] That is, each doubling in production quantity results in an 8 percent unit cost decrease.

[9] The B-kit and A-kit weight estimates are based on weights provided for comparable military systems (from Popp, 2003) adjusted to fit within the volumetric constraints on the underside of a representative commercial aircraft.

- an assumed duty cycle of 55 percent based on the system being on for only 30 minutes during takeoff and 30 minutes during landing for all short-haul and long-haul (international and domestic) commercial aircraft
- a built-in-test or health monitoring subsystem sufficient in capability to reduce the elapsed turnaround time for on-equipment (airport) maintenance to 30 minutes or less for short-haul flights and between two and four hours for long-haul international or domestic flights
- finally, another 25 percent added to the maintenance and delay costs ($40,000 per aircraft per year) to account for the cost growth uncertainty in the estimates, since there is only limited field experience from which to infer O&S costs, and that is with a military DIRCM system based on a lamp, not a laser; the assumed 800-hour MTBF is the reliability derived from the most analogous LAIRCM system, which is based on lower-level hardware estimates

The total O&S cost is thus $27 billion. Once all systems are installed, annual O&S costs would amount to a little over $300,000 per airplane, or $2.1 billion for a 6,800-plane fleet.

Two O&S cost-related issues could add to the overall uncertainty and potential cost growth of the LCC estimates. First, how would passenger flight safety concerns related to a faulty airborne DIRCM system affect the airline industry's record for on-time departures, if the countermeasure system were considered flight-critical hardware? What will be the criteria by which airline mechanics and airport schedulers decide it is prudent to delay scheduled departures? Decisions for this system are more comparable to a breach or malfunction in the security doors of the pilot's cockpit as opposed to detecting an oil leak coming from one of the engines. Second, our LCC estimates are based on designing and producing robust, highly reliable systems that will allow for on-equipment airport turnaround times fast enough to fit within most of today's flight schedules. MTBFs that exceed our assumed values or failure of the built-in test system to allow rapid enough diagnostics could lead to

- procuring higher quantities of spares (initially and annually) to supply airport maintenance activities
- late departures or flight cancellations, as noted above

Our uncertainty allowance may cover the first of these consequences, but it was not intended to encompass the second. The costs of late departures could be substantial. For example, suppose all detected DIRCM system failures take more than 30 minutes to fix for short-haul flights, and 75 percent of them take more than four hours for long-haul flights. If the net revenue loss for each hour of delayed departure is $10,000, the annual O&S cost would increase by 18 percent, from $2.1 billion to $2.5 billion. To avoid such losses, airlines may choose to increase the use of available aircraft at airports or activate reserve aircraft to fill in, but these options have their own costs.

One of the goals of a countermeasures development program would be to increase the reliability of such systems in order to reduce the O&S costs.[10] We examined the impact of in-

[10] See S. Erwin, "Anti-Missile Program for Airliners on a Fast Track," *National Defense*, December 2003, http://www.nationaldefensemagazine.org/article.cfm?Id=1281 (as of February 1, 2004).

creasing the MTBF of countermeasure systems to 3,000 hours (vice 800). Annual O&S costs fell from $2.1 billion to $0.9 billion, and ten-year LCCs fell from $38 billion to $25 billion, so there is a significant potential payoff involved in increasing the installation costs from $11.2 billion to $13.4 billion for designing and procuring a more reliable system.

There is one further element we have not included. Even though the commercial DIRCM system will be designed to operate autonomously and not require pilot intervention, it will need to include communications links for transmitting data directly to law enforcement, transportation security, aviation safety, and homeland-security authorities located at ground command and control ($C^2$) centers. In other words, it needs to be able to integrate into an overall aviation safety and security *system*. The purpose of this link into the system is threefold. First, the link would serve to pass system reliability information on to aviation security officials. This would inform flight procedures if countermeasure systems break down prior to takeoff or during flight. Second, the link should inform aviation safety officials when MANPADS-related events are being detected so as to notify nearby aviation traffic of possible danger. Third, the link should inform law enforcement authorities by indicating the nature of the attack and the estimated location(s) of attackers. It should be noted that the latter two could include false alarms, so procedures need to properly balance security, safety, and economic issues. Even though the cost of this communications link is included in the installation cost estimate, the entire *system* concept of operations should be articulated clearly prior to finalizing system design and before starting LRIP. In addition, the cost of the infrastructure, staff, and equipment for ground-based $C^2$ centers would have to be added to the total LCC estimate as part of the overall systems architecture.

Even if the total development, procurement, and installation costs, estimated at $11 billion, were fully borne by the federal government, this expense covers only 29 percent of the total LCC of $38 billion. There is no guarantee that the government will pay all or any of the $27 billion costs of operating and supporting the airborne DIRCM systems from FY 2007 through FY 2023. Even in the highly competitive environment of the U.S. commercial-airline industry, this additional O&S expense and potential loss of revenue can only be made up through increasing passenger ticket prices.

Given the magnitude of the uncertainty described above and other related factors that are projected to drive maintenance costs, while the O&S cost could certainly increase, improved reliability of the system would reduce these costs by a factor of two or more. Therefore, one of the primary objectives of any countermeasure development and evaluation effort should be to reduce those uncertainties.

## Budgetary Considerations

Clearly, MANPADS countermeasure implementation will be costly. While the potential economic and strategic costs stemming from an attack could be even greater, allocation of resources to countermeasures may mean a reduction in resources applied to other parts of aviation security, as well as to other homeland-security and counterterrorism efforts. More broadly, resource allocation for protecting commercial airliners should strive to be based upon risk, vulnerability, cost, and benefit. They would weigh the risk from a variety of attacks, such as MANPADS, bombs, small arms, and RPGs and then compare side by side the cost and benefits of various counters, such as IRCMs, bomb-resistant containers, and airport security procedures.

Any decision about government-mandated countermeasures installation aboard commercial airliners should thus consider the overall budget available (or necessary) for homeland-security purposes and the more general struggle against terrorism. The FY 2004 Department of Homeland Security (DHS) budget is $36.5 billion. In the broader struggle against terrorism, the President's budget request for FY 2004 included $16 billion for military operations and recovery efforts in Afghanistan and $71 billion for Iraq.[11] In comparison to these figures, the anticipated $2.1 billion annual O&S cost for MANPADS countermeasures seems small. However, as indicated by the DHS budget breakdown in Figure 7.1, $2.1 billion is a substantial fraction (almost half) of the resources being devoted to all of transportation security in the United States today (the pie slice labeled *TSA*). And countermeasures against MANPADS are only one layer of responses against one of many possible threats to air travel.

We should note that our cost estimates encompass only the period through 2023. At that point or even earlier, it may become desirable to replace the laser-jammer systems with HELs, if more sophisticated MANPADS proliferate among terrorists. In contrast to the next several years, in which countermeasure installation can begin with no preexisting O&S demand, the installation costs of future countermeasure generations could lead to total amortized program costs well in excess of the annual O&S figure we cite here.

**Figure 7.1**
**FY 2004 Expenditures for DHS (billions of dollars)**

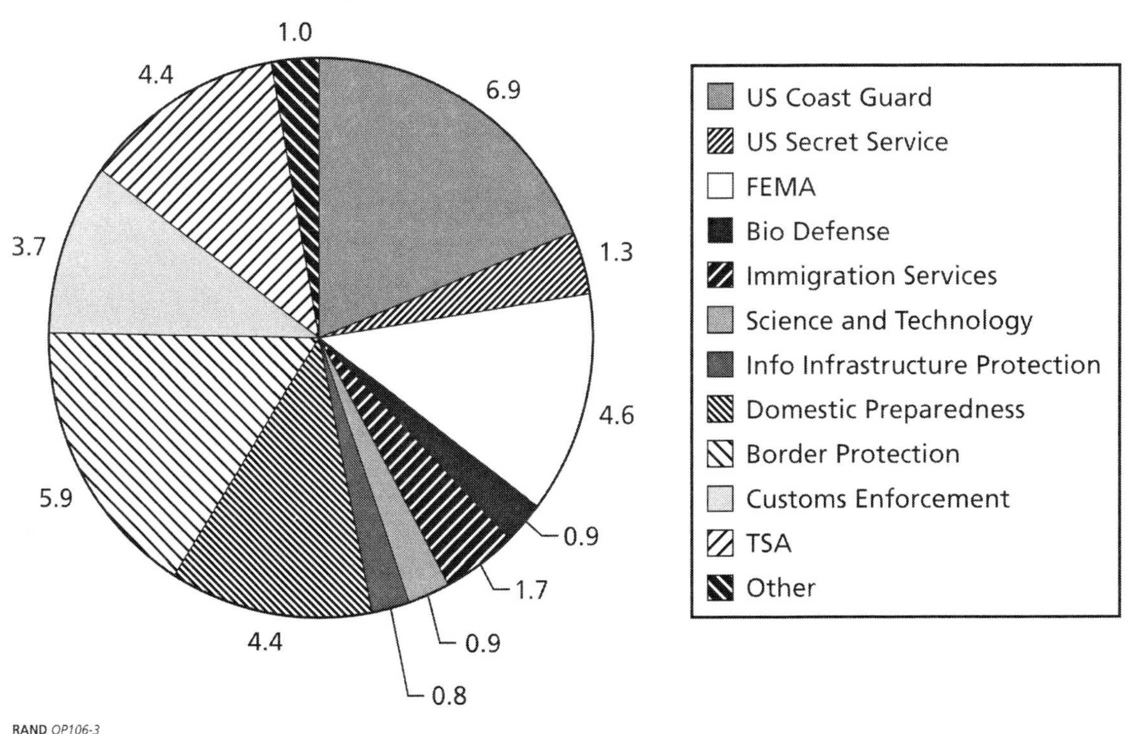


RAND *OP106-3*

---

[11] Data for homeland-security expenditure comes from Department of Homeland Security, "FY2004 Budget Fact Sheet," http://www.dhs.gov/dhspublic/display?content=1817 (as of October 2003). The spending estimate for the military and reconstruction efforts in Afghanistan and Iraq come from the President's FY 2004 supplemental appropriations budget request.

CHAPTER EIGHT

# Summary and Recommendations

Air travel has become an integral part of modern life. Terrorists have long understood this and have made commercial aviation one of their prime targets. Al Qaeda and its affiliates have both the motive and the means to bring down U.S. commercial aircraft with MANPADS. No such attempt has yet been made against a U.S. carrier, but given the measures being taken to preclude 9/11-style attacks, the use of MANPADS will unavoidably become more attractive to terrorists.

What might be done to prevent such an attack? We concentrate here on the capabilities and costs of onboard technologies to divert or destroy an attacking missile. Given the significant costs involved with operating countermeasures based upon current technology, we believe a decision to install such systems aboard commercial airliners should be postponed until the technologies can be developed and shown to be more compatible in a commercial environment. This development effort should proceed as rapidly as possible. Concurrently, a development effort should begin immediately that focuses on understanding damage mechanisms and the likelihood of catastrophic damage to airliners from MANPADS and other forms of man-portable weapons. Findings from the two development programs should inform a decision on the number of aircraft that should be equipped with countermeasures (from none to all 6,800 U.S. jet-powered airliners) and the sequence in which aircraft are to be protected.

If it is determined that U.S. commercial airliners should be equipped with countermeasures upon completion of the development program, they should be employed as part of a broader set of initiatives aimed at striking and capturing terrorists abroad, impeding their acquisition of missiles, and preventing them and their weapons from entering the United States. Attention should also be paid to keeping MANPADS-equipped terrorists out of areas adjacent to airports and improving commercial airliners' ability to survive fire-induced MANPADS damage.

A multilayered approach is important because no single countermeasure technology can defeat all possible MANPADS attacks with high confidence. Nonetheless, substantial protection can be achieved. Laser jammers, for instance, will be commercially available for installation aboard airliners soon and should be able to divert single or possibly dual attacks by the relatively unsophisticated MANPADS accounting for most of those now in the hands of terrorists. Ground-based HELs intended to destroy approaching missiles could counter MANPADS of any degree of sophistication, but they are not ready for deployment in the next few years and have significant operational challenges to overcome. Pyrophoric flares used reactively offer the promise of a cheaper alternative with better potential to handle multiple attacks than laser-based systems, but their effectiveness at protecting large transport aircraft from any

MANPADS attack is not well established, and they would be most likely ineffective against sophisticated future systems.

We estimate that it would cost about $11 billion to install a single laser jammer on each of the 6,800 commercial aircraft in the U.S. fleet. The operating costs of fleetwide countermeasures will depend on the reliability of the system. Extrapolating from early reliability data from the systems currently deployed on large military aircraft, the O&S costs for a commercial variant were assessed to be $2.1 billion per year for the entire commercial fleet. The full ten-year LCCs for developing, installing, operating, and supporting laser-jammer countermeasures are estimated to be $40 billion. If reliability goals recommended by DHS can be achieved, the ten-year LCCs are estimated to be $25 billion.

When would such an investment be worth it? That is not a question answerable solely through quantitative analysis, but some light can be shed by four avenues of inquiry. First, what would be the likely economic costs of a successful attack? If we take into account the value of a lost aircraft and a conventional economic valuation of loss of life, the *direct* cost would approach $1 billion for every aircraft downed. The indirect economic damage from an attack would be far greater. These costs result from the loss of consumer welfare through preemption of a favored travel mode or reluctance to use it, as well as operating losses suffered by airlines subsequent to an attack. These amounts will depend primarily upon two factors: the length of any possible systemwide shutdowns in air travel and any kind of longer-lasting public reluctance to fly. Both factors are difficult to predict, but if air travel were shut down for a week (it was shut down for three days after 9/11), the economic loss would amount to roughly $3 billion during the shutdown itself. Extrapolating from the long-term effects of the 9/11 shutdown, losses over the following months might tally an additional $12 billion, for a total economic impact of more than $15 billion.

A second avenue of inquiry can help place the cost of MANPADS countermeasures in context. To what extent must homeland-security and other counterterrorism resources be expanded or diverted to fund this one effort to help respond to a single threat? The $2.1 billion annual O&S cost, should it be borne by the government, amounts to only about 6 percent of the annual DHS budget. The fraction is much smaller if the costs of operations in Iraq and Afghanistan are included in the base. However, the $2.1 billion is a substantial fraction of total current federal expenditures on transportation security.

Third, it must be recognized that loss of life and economic impact would not be the only costs of a MANPADS attack. The perceived inability of the U.S. government to prevent attacks on its citizens on its own soil would set back U.S. efforts to counter terrorist groups globally and could weaken U.S. influence across a range of other interests abroad. Such an attack would also cause unquantifiable losses of security among the U.S. populace.

Fourth, and lastly, while countermeasures have been demonstrated to be an effective resource in protecting our military aircraft, the circumstances of protecting commercial airliners from terrorists are sufficiently different that we should ask ourselves the following questions: Upon deployment of countermeasures, how easy do we think it will be for terrorists to adapt and find vulnerabilities to airliners through the use of weapons that are not affected by countermeasures? Would defenses against these weapons be possible, or would they require a similar level of funding to protect against?

A decision as to whether to proceed with a MANPADS countermeasure program must thus balance a variety of considerations. On the plus side:

- New countermeasure technology with capability against a variety of attack situations will be available in the near term, with the potential to avert the loss of hundreds or even thousands of lives and tens of billions of dollars.
- Funding such a system would require a reallocation or expansion of federal homeland-security resources of perhaps 5 percent—and a much smaller proportion of total federal counterterrorism resources.

On the minus side:

- Annual operating costs would represent nearly 50 percent of what the federal government currently spends for all transportation security in the United States.
- Well-financed terrorists will likely always be able to devise a MANPADS attack scenario that will defeat whatever countermeasures have been installed, although countermeasures can make such attacks considerably more difficult and less frequent.
- Installing countermeasures to MANPADS attacks may simply divert terrorist efforts to less protected opportunities for attack. To put it another way, how many avenues for terrorist attack are there, and can the United States afford to block them all?

Given the significant uncertainties in the cost of countermeasures and their effectiveness in reducing our overall vulnerability to catastrophic airliner damage, a decision to install should be postponed, and concurrent development efforts focused on reducing these uncertainties should proceed as rapidly as possible. The current DHS RDT&E activities are a prudent step both toward reducing significant cost uncertainties involved and minimizing the delay of program implementation once a go-ahead decision is reached.

To summarize, any federal policy to protect against MANPADS should not be restricted to countermeasures development but should involve multiple layers, with emphasis on the following areas:

1. Rapidly understanding and finding ways to reduce the O&S cost component of countermeasures in a commercial-airline setting. In addition, decisionmakers should be thinking about how specific countermeasure systems would work best in conjunction with other protection efforts and technologies. Understanding the weaknesses of countermeasures should help focus these efforts, and vice versa.
2. Focusing a concurrent technology development effort on understanding damage mechanisms and the likelihood of catastrophic damage to airliners from MANPADS and other forms of man-portable weapons such as RPGs, mortars, and small-arms fire. This will serve three purposes: clarifying the damage caused by single or multiple MANPADS hits on airliners, informing choices regarding the implementation of mitigating measures such as inerting fuel tanks and missile countermeasure systems, and assessing the seriousness of other forms of attack against airliners.

3. Working with international governments to slow down the proliferation of MANPADS technologies, in particular those against which countermeasures are less effective.
4. Putting together concepts of operation that integrate countermeasures into the overall aviation safety, security, and law enforcement system. These can help local law enforcement establish the size and location of airport security perimeters and define ways in which information from the onboard countermeasure system sensors can be used to help find, track, and apprehend MANPADS operators. Lastly, they would help provide an understanding of the costs from false alarms to air-traffic operations and local law enforcement.

APPENDIX A

# Estimating Consumer Surplus Loss

To estimate the actual consumer surplus for a successful shoot-down of a commercial aircraft in the United States, we first divide air travel into different market segments based on length of trip. Then we estimate the cost of travel for different travel modes, including air travel and its alternatives, in these market segments. We examine travelers' willingness to pay to avoid the shutdown of air travel, which varies by segment, and calculate the consumer surplus loss that results.

**Segmenting the market.** Table A.1 provides an estimate of the number of household trips taken by commercial airplane—an estimate that is used to calculate the percentage share of air-traveler miles according to the round-trip distance traveled for each trip. These trips are sorted into five different distance categories. An average number of miles flown is taken for each distance category and multiplied by the number of trips to generate the number of miles flown for each distance. This calculation generates the distribution of miles among different trip distances, which will be used in later calculations. The vast majority of airline passenger miles occurs during long trips—those at least 1,000 miles in each direction.

We break the market for air travel down further according to trip purpose (business or leisure) and destination (domestic or international). Business and leisure travelers tend to make different decisions about the speed and cost of transportation, and they tend to value time differently. On average, half of airline travel tends to be business and the other half leisure. Domestic and international travel is split because relatively convenient alternatives exist for domestic flights, while without air travel international trips to any place outside of Canada and Mexico become highly difficult. For Americans, about three-quarters of air mileage is domestic, and the rest international.

**Estimating the costs of alternatives.** In the case of an air-travel shutdown, travelers would need to take a car, train, or bus to reach their destination; some travel would undoubtedly be canceled. More than 90 percent of all non-air trips, even at the longest distances, are taken by car. Furthermore, transportation-mode-choice models of intercity travel generally show bus travel to have a specific disutility associated with it that cannot be easily attributed to its cost or its speed.[1] That is, many travelers exhibit distaste for bus travel that cannot be readily translated to the kind of welfare calculations we are going to make. Therefore, we will consider car and train as the two alternatives when air travel is disrupted.

We next estimate the change in trip cost that would occur when switching from flying to using an alternative mode. For international travel outside of North America, which means

[1] For example, see Steven A. Morrison and Clifford Winston, "An Econometric Analysis of the Demand for Intercity Passenger Transportation," *Research in Transportation Economics*, Vol. 2, 1985, pp. 213–37.

**Table A.1**
**Aviation Miles by Trip Distance**

| Round Trip Distance | Average Round Trip Distance | Commercial Aviation Trips (thousands) | Commercial Aviation Miles (thousands) | Percent of Total Miles |
|---|---|---|---|---|
| <300 miles | 250 | 1,364 | 341,000 | 0.1% |
| 300–499 | 425 | 7,118 | 3,025,150 | 1.1% |
| 500–999 | 800 | 26,812 | 21,449,600 | 7.9% |
| 1,000-1,999 | 1600 | 36,294 | 58,070,400 | 21.4% |
| >2,000 miles | 3525 | 53,295 | 187,862,362 | 69.4% |

Source: USDOT (1997)

most international flights, no viable alternative mode exists. For domestic trips, we calculate the time and cost involved for each of the three modes: air, car, and rail, for each average round-trip distance (see Table A.2). Travel times are calculated based on travel speeds. The trip distance is divided by the average speed of each mode to generate a trip time in hours. Two additions are made to these travel-time calculations. For air travel, time is added for travel to and from the airport and for the time required for check-in and security screening. For car travel, allowance needs to be made for trips that would last longer than a normal driving day. To translate trip times to dollar values, we use estimates of the value of travel time for business and leisure travelers. The value of time for leisure travelers is taken as $19.50 per hour, and the value of time for business travelers is given as $34.50 per hour, per FAA guidance.[2]

Cost data for air travel comes from the Air Transport Association (http://www.airlines.org/). Passenger yields in 2001 for domestic air travel averaged 13.4 cents per mile. Since exact cost data are not available for passenger operations separate from cargo operations, we assume that per-passenger revenues are approximately equal to per-passenger costs. Auto-travel costs are calculated by the consulting firm of Runzheimer International and are available at their Web site (http://www.runzheimer.com/). The full costs of owning and operating an automobile were calculated at 52 cents per mile. However, on average most auto users do not travel alone. Therefore, we assume that two people take the average auto trip and thus figure the cost at 26 cents per mile. Rail-travel costs come from the Amtrak Annual Report. The 25.7 cents per-mile cost used here is for a seat mile of travel.

When time and travel costs are added, we find that for trips of 800 miles and less, the sum is lower for car travel than for train travel. However, train travel is less costly than car travel for trips in the longest two distance categories. This holds true for both business and leisure travel. The combined cost difference between air travel and the best alternative is calculated and expressed as both a percentage change and a per-mile difference; this is then used in calculating the change in consumer welfare.

**Calculating consumer surplus.** To estimate consumer surplus, we need to know the difference between consumer willingness to pay and price. For this study, we approximate the willingness to pay for air travel by using the elasticity of demand. The elasticity of demand describes the percentage change in demand for air travel given a percentage change in its price.

[2] FAA, 1998.

**Table A.2**
**Travel Costs for Air and Alternative Modes**

| | Leisure Travel | | | | |
|---|---|---|---|---|---|
| **Airline** | **Cost per mile: $0.134** | **Time cost: $19.50** | | **Avg. Speed: 400** | |
| | **Trip Length** | **Time (hours)** | **Time Cost** | **Travel Cost** | **Total Cost** |
| | 250 | 6.63 | $129 | $34 | $163 |
| | 425 | 7.06 | $138 | $57 | $195 |
| | 800 | 8.00 | $156 | $107 | $263 |
| | 1600 | 10.00 | $195 | $214 | $409 |
| | 3525 | 14.81 | $289 | $472 | $761 |
| **Car** | **Cost per mile: $0.260** | **Avg. Speed: 60** | | | |
| | **Trip Length** | **Time (hours)** | **Time Cost** | **Travel Cost** | **Total Cost** |
| | 250 | 4.17 | $81 | $65 | $146 |
| | 425 | 7.08 | $138 | $111 | $249 |
| | 800 | 13.33 | $260 | $208 | $468 |
| | 1600 | 42.67 | $832 | $416 | $1248 |
| | 3525 | 114.75 | $2238 | $917 | $3154 |
| **Amtrak** | **Cost per mile:$0.257** | **Avg. Speed: 45** | | | |
| | **Trip Length** | **Time (hours)** | **Time Cost** | **Travel Cost** | **Total Cost** |
| | 250 | 5.56 | $108 | $64 | $173 |
| | 425 | 9.44 | $184 | $109 | $293 |
| | 800 | 17.78 | $347 | $206 | $552 |
| | 1600 | 35.56 | $693 | $411 | $1105 |
| | 3525 | 78.33 | $1528 | $906 | $2433 |

| Best Alternative | Trip Length | Total Cost Change | Total Cost Change Pct. | Total Cost Change/Mile |
|---|---|---|---|---|
| Car | 250 | –$16 | –10% | $0.066 |
| Car | 425 | $54 | 28% | $0.127 |
| Car | 800 | $205 | 78% | $0.256 |
| Amtrak | 1600 | $695 | 170% | $0.434 |
| Amtrak | 3525 | $1672 | 220% | $0.474 |

With an elasticity of –1, for example, for every 10 percent increase in the price of travel, 10 percent fewer people will be willing to pay to make the trip. Put the other way, 90 percent of people would value air travel enough to pay 10 percent more to still make the trip.

To estimate the change in consumer surplus when air travel is disrupted, we first calculate how many people paid what price when they were able to fly, and then we calculate how many people would be willing to pay the price of taking the best alternative mode of travel. We compare how much higher the price of the alternative is and how many people decided to cancel their trip. Those that cancel have lower values of making the trip and lose the consumer surplus they would have received from their travel. Those that still travel by the alternative lose consumer surplus from the higher price they have to pay.[3]

[3] In estimating the number of travelers who would still take the trip using an alternative mode, we assume that tastes for travel and income levels do not change, neither of which may hold true in the event of a catastrophic terrorist attack. An attack could also affect the preference between car and train for travel in the aftermath. However, the nature and size of the effect is uncertain a priori and is not included here.

**Table A.2**
**(continued)**

| Business Travel | | | | | |
|---|---|---|---|---|---|
| **Airline** | **Time cost: $34.50** | **Cost per mile: $0.134** | **Avg. Speed: 400** | | |
| | **Trip Length** | **Time (hours)** | **Time Cost** | **Travel Cost** | **Total Cost** |
| | 250 | 6.63 | $229 | $34 | $262 |
| | 425 | 7.06 | $244 | $57 | $301 |
| | 800 | 8.00 | $276 | $107 | $383 |
| | 1600 | 10.00 | $345 | $214 | $559 |
| | 3525 | 14.81 | $511 | $472 | $983 |
| **Car** | **Cost per mile: $0.260** | **Avg. Speed: 60** | | | |
| | **Trip Length** | **Time (hours)** | **Time Cost** | **Travel Cost** | **Total Cost** |
| | 250 | 4.17 | $144 | $65 | $209 |
| | 425 | 7.08 | $244 | $111 | $355 |
| | 800 | 13.33 | $460 | $208 | $668 |
| | 1600 | 42.67 | $1472 | $416 | $1888 |
| | 3525 | 114.75 | $3959 | $917 | $4875 |
| **Amtrak** | **Cost per mile: $0.257** | **Avg. Speed: 45** | | | |
| | **Trip Length** | **Time (hours)** | **Time Cost** | **Travel Cost** | **Total Cost** |
| | 250 | 5.56 | $192 | $64 | $256 |
| | 425 | 9.44 | $326 | $109 | $435 |
| | 800 | 17.78 | $613 | $206 | $819 |
| | 1600 | 35.56 | $1227 | $411 | $1638 |
| | 3525 | 78.33 | $2703 | $906 | $3608 |

| Best Alternative | Trip Length | Total Cost Change | Total Cost Change Pct. | Total Cost Change/Mile |
|---|---|---|---|---|
| Car | 250 | –$53 | –20% | –$0.213 |
| Car | 425 | $54 | 18% | $0.356 |
| Car | 800 | $285 | 74% | $0.128 |
| Amtrak | 1600 | $1078 | 193% | $0.674 |
| Amtrak | 3525 | $2625 | 267% | $0.745 |

SOURCES: Runzheimer International, Amtrak, Air Transport Association

In Table A.3, to calculate the consumer welfare, the trip-distance mileage-share calculations from Table A.1 are first applied to revenue passenger mile data. This generates the number of miles traveled for each trip distance and purpose category. Price elasticities of demand for business and leisure travelers are used to translate the cost changes for the best travel alternative from Table A.2 into welfare changes. Welfare changes are then calculated for a systemwide shutdown of a day, a week, and a month.

As mentioned previously, most international air travel does not have a realistic alternative. Accordingly, we assume a price increase for air travel large enough to preclude essentially all demand. Welfare changes are also calculated for international travel. Using this approach, combined consumer welfare losses from a month-long shutdown of domestic and international flights would top $8 billion, while a week's shutdown would incur about $2 billion worth of lost travel value.

**Table A.3**
**Consumer Surplus Loss (Domestic)**
**(all miles and dollars in billions, except cost change in dollars per mile)**

**Domestic**

**Revenue Passenger Miles (2001):** 480 **Business/Leisure split:** 50%
**Business Miles:** 240 **Price Elasticity of Demand:** –0.7

| RT Mileage | Precentage of Total Miles | Miles | Total Cost Change per Mile | Travel Cost Change Pct. | Shutdown Cost of One Day | Shutdown Cost of One Week | Shutdown Cost of One Month |
|---|---|---|---|---|---|---|---|
| <300 | 0.1% | 0.3 | –$0.213 | –20% | $0.00 | $0.00 | –$0.01 |
| 300–499 | 1.1% | 2.7 | $0.128 | 18% | $0.00 | $0.01 | $0.03 |
| 500–999 | 7.9% | 19.0 | $0.356 | 74% | $0.01 | $0.10 | $0.41 |
| 1,000-1,999 | 21.4% | 51.4 | $0.674 | 193% | $0.04 | $0.25 | $1.06 |
| >2,000 | 69.4% | 166.4 | $0.745 | 267% | $0.09 | $0.64 | $2.72 |
| **Domestic business subtotal** | | | | | $0.14 | $0.98 | $4.21 |

**Leisure miles:** 240 **Price elasticity of demand:** –1.0

| RT Mileage | Percentage of Total Miles | Miles | Total Cost Change per Mile | Travel Cost Change Pct. | Shutdown Cost of One Day | Shutdown Cost of One Week | Shutdown Cost of One Month |
|---|---|---|---|---|---|---|---|
| <300 | 0.1% | 0.3 | –$0.066 | –10% | $0.00 | $0.00 | $0.00 |
| 300-499 | 1.1% | 2.7 | $0.127 | 28% | $0.00 | $0.01 | $0.02 |
| 500-999 | 7.9% | 19.0 | $0.256 | 78% | $0.01 | $0.06 | $0.24 |
| 1,000-1,999 | 21.4% | 51.4 | $0.434 | 170% | $0.02 | $0.13 | $0.54 |
| >2,000 | 69.4% | 166.4 | $0.474 | 220% | $0.05 | $0.34 | $1.48 |
| **Domestic leisure subtotal** | | | | | $0.08 | $0.53 | $2.28 |
| **Domestic subtotal** | | | | | $0.22 | $1.52 | $6.50 |

**International**

**Revenue Passenger Miles (2001): 175.8 Business/Leisure split: 50%**
**Business Price Elasticity: –0.7 Leisure Price Elasticity: –1.0**

| | Miles | Total Cost Change per Mile | Travel Cost Change Pct. | Shutdown Cost of One Day | Shutdown Cost of One Week | Shutdown Cost of One Month |
|---|---|---|---|---|---|---|
| **Business miles** | 88 | $0.346 | 143% | $0.04 | $0.29 | $1.27 |
| **Leisure miles** | 88 | $0.179 | 100% | $0.02 | $0.15 | $0.66 |
| **International subtotal** | | | | $0.06 | $0.44 | $1.92 |
| **Domestic and international subtotal** | | | | $0.28 | $1.96 | $8.42 |

SOURCE: Air Transport Association

**Table A.4**
**Producer Surplus Loss**
**(all miles and dollars in billions, except revenue per passenger mile in dollars per mile)**

| | | System Shutdown of | |
|---|---|---|---|
| **Domestic** | **One Day** | **One Week** | **One Month** |
| Revenue Passenger Miles (2001) | 1.31 | 9.20 | 39.41 |
| Revenue per Passenger Mile (2001) | $0.134 | $0.134 | $0.134 |
| Domestic Airline Revenue | $0.2 | $1.2 | $5.3 |
| | | **System Shutdown of** | |
| **International** | **One Day** | **One Week** | **One Month** |
| Revenue Passenger Miles (2001) | 0.48 | 3.37 | 14.45 |
| Revenue per Passenger Mile (2001) | $0.097 | $0.097 | $0.097 |
| International Airline Revenue | $0.0 | $0.3 | $1.4 |
| **Subtotal** | $0.22 | $1.56 | $6.68 |
| **Percent of costs incurred** | 87% | 87% | 84% |
| **Total** | $0.19 | $1.36 | $5.61 |

SOURCE: Air Transport Association

For consistency, Table A.4 shows producer surpluses in a similar format. Combining the numbers from Tables A.3 and A.4, the sum of consumer and producer surpluses tops $3 billion for a one-week shutdown and comes to $14 billion for one month.

# Congressional Bills

Listed are two relevant congressional bills introduced during 2003 and 2004 dealing with countermeasures installation aboard commercial airliners.

HR 4056 IH 108th CONGRESS 2d Session, **H. R. 4056**

To encourage the establishment of both long-term and short-term programs to address the threat of man-portable air defense systems (MANPADS) to commercial aviation.

**IN THE HOUSE OF REPRESENTATIVES**

March 30, 2004

Mr. MICA (for himself, Mr. DEFAZIO, and Mr. ISRAEL) introduced the following bill; which was referred to the Committee on Transportation and Infrastructure, and in addition to the Committee on International Relations, for a period to be subsequently determined by the Speaker, in each case for consideration of such provisions as fall within the jurisdiction of the committee concerned.

A bill:

To encourage the establishment of both long-term and short-term programs to address the threat of man-portable air defense systems (MANPADS) to commercial aviation.

*Be it enacted by the Senate and House of Representatives of the United States of America in Congress assembled,*

SECTION 1. SHORT TITLE.

This Act may be cited as the 'Commercial Aviation MANPADS Defense Act of 2004'.

SEC. 2. FINDINGS.

Congress finds the following:

(1) MANPADS constitute a threat to military and civilian aircraft.

(2) The threat posed by MANPADS requires the development of both short-term and long-term plans.

(3) The threat posed by MANPADS requires an international as well as domestic response.

(4) There should be an international effort to address the issues of MANPADS proliferation and defense.

(5) The Government is pursuing and should continue to pursue diplomatic efforts to prevent the proliferation of MANPADS.

SEC. 3. INTERNATIONAL COOPERATIVE EFFORTS.

(a) To Limit Availability and Transfer of MANPADS- The President is encouraged to pursue further strong international diplomatic and cooperative efforts, including bilateral and multilateral treaties, in the appropriate forum to limit the availability, transfer, and proliferation of MANPADS worldwide.

(b) To Achieve Destruction of MANPADS- The President should continue to pursue further strong international diplomatic and cooperative efforts, including bilateral and multilateral treaties, in the appropriate forum to assure the destruction of excess, obsolete, and illicit stocks of MANPADS worldwide.

(c) Reporting and Briefing Requirements- Not later than 180 days after the date of enactment of this Act, the President shall transmit to the appropriate congressional committees a report that contains a detailed description of the status of diplomatic efforts under subsections (a) and (b). Annually thereafter until completion of such diplomatic efforts, the Secretary of State shall brief the appropriate congressional committees on the status of such diplomatic efforts.

SEC. 4. FAA AIRWORTHINESS CERTIFICATION OF MISSILE DEFENSE SYSTEMS FOR COMMERCIAL AIRCRAFT.

(a) In General- Not later than 30 days after the date of enactment of this Act, the Administrator of the Federal Aviation Administration shall establish a process for conducting airworthiness and safety certification of missile defense systems for commercial aircraft.

(b) Certification Acceptance- Under the process, the Administrator shall accept the certification of the Department of Homeland Security that a missile defense system is effective and does not pose a danger when used to defend commercial aircraft against MANPADS.

(c) Expeditious Certification- Under the process, the Administrator shall expedite the airworthiness and safety certification of missile defense systems for commercial aircraft.

(d) Reports- Not later than 180 days after the initiation of certification procedures for missile defense systems for commercial aircraft, and every 6 months thereafter until complete, the Federal Aviation Administration shall transmit to the Committee on Transportation and Infrastructure of the House of Representatives and the Committee on Commerce, Science, and Transportation of the Senate a report that contains a detailed description of the status of airworthiness and safety certification.

SEC. 5. PROGRAMS TO REDUCE MANPADS.

(a) In General- The President is encouraged to pursue strong programs to reduce the number of MANPADS worldwide so that fewer MANPADS will be available for trade, proliferation, and sale.

(b) Reporting and Briefing Requirements- Not later than 180 days after the date of enactment of this Act, the President shall transmit to the appropriate congressional committees a report that contains a detailed description of the status of the programs being pursued under subsection (a). Annually thereafter until the programs are no longer needed, the Secretary of State shall brief the appropriate congressional committees on the status of programs.

(c) Funding- There is authorized to be appropriated such sums as may be necessary to carry out this section.

SEC. 6. MANPADS VULNERABILITY ASSESSMENTS REPORT.

(a) In General- Not later than one year after the date of enactment of this Act, the Secretary of Homeland Security shall transmit to the Committee on Transportation and Infrastructure of the House of Representatives and the Committee on Commerce, Science, and Transportation of the Senate a report describing the Department of Homeland Security's plans to secure airports and the aircraft arriving and departing from airports against MANPADS attacks.

(b) Matters to Be Addressed- The Secretary's report shall address, at a minimum, the following:

(1) The status of the Department's efforts to conduct MANPADS vulnerability assessments at United States airports at which the Department is conducting assessments.

(2) How intelligence is shared between the United States intelligence agencies and Federal, State, and local law enforcement to address the MANPADS threat and potential ways to improve such intelligence sharing.

(3) Contingency plans that the Department has developed in the event that it receives intelligence indicating a high threat of MANPADS attack on aircraft at or near United States airports.

(4) The feasibility and effectiveness of implementing public education and neighborhood watch programs in areas surrounding United States airports in cases in which intelligence reports indicate there is a high risk of MANPADS attacks on aircraft.

(5) Any other issues that the Secretary deems relevant.

(c) Format- The report required by this section may be submitted in a classified format.

SEC. 7. DEFINITIONS.

In this Act, the following definitions apply:

(1) Appropriate congressional committees- The term 'appropriate congressional committees' means—

(A) the Committee on Armed Services, the Committee on International Relations, and the Committee on Transportation and Infrastructure of the House of Representatives; and

(B) the Committee on Armed Services, the Committee on Foreign Relations, and the Committee on Commerce, Science, and Transportation of the Senate.

(2) MANPADS- The term 'MANPADS' means man-portable air defense systems, which are shoulder-fired, surface-to-air missile systems that can be carried and transported by a person.

HR 580 IH 108th CONGRESS 1st Session, **H. R. 580**

To direct the Secretary of Transportation to issue regulations requiring turbojet aircraft of air carriers to be equipped with missile defense systems, and for other purposes.

**IN THE HOUSE OF REPRESENTATIVES**

February 5, 2003

Mr. ISRAEL introduced the following bill; which was referred to the Committee on Transportation and Infrastructure, and in addition to the Committee on Armed Services, for a period to be subsequently determined by the Speaker, in each case for consideration of such provisions as fall within the jurisdiction of the committee concerned.

A bill:

To direct the Secretary of Transportation to issue regulations requiring turbojet aircraft of air carriers to be equipped with missile defense systems, and for other purposes.

*Be it enacted by the Senate and House of Representatives of the United States of America in Congress assembled,*

SECTION 1. SHORT TITLE.

This Act may be cited as the 'Commercial Airline Missile Defense Act'.

SEC. 2. REGULATIONS REQUIRING MISSILE DEFENSE SYSTEMS.

(a) IN GENERAL- Not later than 90 days after the date of enactment of this Act, the Secretary of Transportation shall issue regulations that require all turbojet aircraft used by an air carrier for scheduled air service to be equipped with a missile defense system.

(b) SCHEDULE FOR INSTALLATION- The regulations shall establish a schedule for the purchase and installation of such systems on turbojet aircraft currently in service and turbojet aircraft contracted for before the date of issuance of the regulations.

(c) NEW AIRCRAFT- The regulations shall also require that all turbojet aircraft contracted for on or after the date of issuance of the regulations by an air carrier for scheduled air service be equipped with a missile defense system.

(d) DEADLINES FOR COMMENCEMENT OF INSTALLATION- The regulations shall require that installation and operation of missile defense systems under the regulations begin no later than December 31, 2003.

SEC. 3. PURCHASE OF MISSILE DEFENSE SYSTEMS BY THE SECRETARY.

The Secretary of Transportation shall purchase and make available to an air carrier such missile defense systems as may be necessary for the air carrier to comply with the regulations issued under section 2 (other than subsection (c)) with respect to turbojet aircraft used by the air carrier for scheduled air service.

SEC. 4. RESPONSIBILITY OF AIR CARRIER.

Under the regulations issued under section 2, an air carrier shall be responsible for installing and operating a missile defense system purchased and made available by the Secretary of Transportation under section 3.

SEC. 5. PROGRESS REPORTS.

Not later than January 1, 2004, and each July 1 and January 1 thereafter, the Secretary of Transportation shall transmit to Congress a report on the progress being made in implementation of this Act, including the regulations issued to carry out this Act.

SEC. 6. INTERIM SECURITY MEASURES

(a) IN GENERAL- In order to provide interim security before the deployment of missile defense systems for turbojet aircraft required under section 2, the President shall—

(1) exercise the President's authority under title 32, United States Code, to elevate National Guard units to Federal status for the purpose of patrolling airport areas surrounding airports against the threat posed by missiles and other ordinance to commercial aircraft; and

(2) deploy units of the United States Coast Guard, in coordination with the Secretary of Transportation and the Secretary of Homeland Security, for the purpose of patrolling areas surrounding airports to protect against the threat posed by missiles and other ordinance to commercial aircraft.

(b) PROGRESS REPORT- Not later than 90 days after the date of enactment of this Act, the President shall submit to Congress a report on the progress being made to implement this section.

SEC. 7. DEFINITIONS.

In this Act, the following definitions apply:

(1) AIRCRAFT AND AIR CARRIER- The terms 'aircraft' and 'air carrier' have the meaning such terms have under section 40102 of title 49, United States Code.

(2) MISSILE DEFENSE SYSTEM- The term 'missile defense system' means an appropriate (as certified by the Secretary of Transportation) electronic system that would automatically—

(A) identify when the aircraft is threatened by an incoming missile or other ordinance;

(B) detect the source of the threat; and

(C) disrupt the guidance system of the incoming missile or other ordinance, which is intended to result in the incoming missile or other ordinance being diverted off course and missing the aircraft.

# References


Air Transport Association, Economics Home Page, June 21, 2003, http://www.airlines.org/econ/p.aspx?nid=6342 (as of November 3, 2003).

———, Airline Industry Facts, Figures, and Analyses, http://www.airlines.org/econ/d.aspx?nid=1026 (as of November 3, 2003).

Benjamin, D., and S. Simon, *The Age of Sacred Terror*, New York: Random House, 2002.

Bolkcom, C., B. Elias, and A. Feickert, *Congressional Research Service Report for Congress: Homeland Security: Protecting Airliners from Terrorist Missiles*, Washington, D.C.: Congressional Research Service 2003.

Caffera, P., "U.S. Jets Easy Target for Shoulder-Fired Missiles," *San Francisco Chronicle,* November 30, 2002, p. A14.

CNN, "Feds Tell How the Weapons Sting Was Played," CNN.com, August 14, 2003, http://cgi.cnn.com/2003/LAW/08/13/arms.sting.details/ (as of November 3, 2003).

CSIS, "Transnational Threats Update," Vol. 1, No. 10, 2003, p.2.

Cullen, Tony, and Christopher F. Foss, eds., *Jane's Air Defense Systems, 2001–2002,* Surrey, England: Jane's Information Group, 2001.

Department of Defense, " FY 2004/2005 President's Budget Item Justification," R-2 RDT&E Exhibit, February 2003.

Department of Homeland Security, "FY2004 Budget Fact Sheet," October 15, 2004, http://www.dhs.gov/dhspublic/display?content=1817 (as of October 2003).

Department of Justice, September 11th Victim Compensation Fund of 2001: Compensation for Deceased Victims, http://www.usdoj.gov/victimcompensation/payments_deceased.html (as of July 19, 2004).

Dixon, L., *Assistance and Compensation for Individuals and Businesses after the September 11th Terrorist Attacks*, Santa Monica, Calif.: RAND Corporation, MG-264-ICJ, forthcoming.

Erwin, S., "Anti-Missile Program for Airliners on a Fast Track," *National Defense,* December 2003, http://www.nationaldefensemagazine.org/article.cfm?Id=1281 (as of February 1, 2004).

Federal Aviation Administration, "Economic Values for Evaluation of FAA Investment and Regulatory Programs," U.S. Department of Transportation, FAA-APO-98-8, June 1998.

Department of the Air Force, *Fiscal Year (FY) 2004/2005 Biennial Budget Estimates, Research, Development, Test and Evaluation (RDT&E), Descriptive Summaries,* Vol. III, Budget Activity 7, February 2003, pp. 1837–1843.

Gusinov, T., "Portable Missiles May Become Next Weapon of Choice for Terrorists," *Washington Diplomat,* June 16, 2003, www.washingtondiplomat.com/03_01/a4_03_01.html (as of March 4, 2004).

Jane's Terrorism and Insurgency Centre, "Proliferation of MANPADS and the Threat to Civil Aviation," August 13, 2003, http://www.janes.com/security/international_security/news/jtic/jtic030813_1_n.shtml (as of September 17, 2003).

Kuhn, D., "Mombassa Attack Highlights Increasing MANPADS Threat," *Jane's Intelligence Review*, February 2003, pp. 26–31.

Morrison, Steven A., and Clifford Winston, "An Econometric Analysis of the Demand for Intercity Passenger Transportation," *Research in Transportation Economics*, Vol. 2, 1985, pp. 213–37.

Office of Management and Budget, "FY-2004 Supplemental Appropriations Request," Washington, D.C., September 17, 2003.

Pedriani, C., "JASPO/NASA Cooperate to Improve Commercial Aviation Security," in *Aircraft Survivability: Reclaiming the Low Altitude Battlespace*, Joint Aircraft Survivability Program Office, 2003.

Popp, M., "Cost Analysis Update," briefing to the Interagency Task Force, NAVAIR 4.2V, February 13, 2003.

Townsend, J., "15 SOS Field Support Visit Aircraft Modernization" unclassified briefing, HQ AFSOC/XPQA, Hurlburt, AFB, January 23, 2001.

Turner, B., "Aviation Demand Forecasts, Large Air Carriers-Passengers, Fiscal Years 2003–2014," FAA, February 13, 2003.

Victoria Transport Policy Institute, "Transporation Costs and Benefits," www.ntpi.org/tdm/tdmbb.htm (as of July 13, 2004).

Wilbur Smith Associates, "The Economic Impact of Civil Aviation on the U.S. Economy," April 2003.